Growing Good Things to Eat in Texas

NUMBER ELEVEN
Texas A&M University Agriculture Series

C. Allan Jones, General Editor

Texas A&M University Press College Station

Growing Good Things to Eat in Texas

Profiles of Organic Farmers and Ranchers across the State

By Pamela Walker

Photographs by Linda Walsh

Manufactured in China by
Everbest Printing Co. through
Four Colour Print Group

First edition

This paper meets the requirements
of ANSI/NISO Z39.48-1992 (Permanence of Paper).
Binding materials have been chosen for durability.

Library of Congress Cataloging-in-Publication Data

Walker, Pamela, 1949–
Growing good things to eat in Texas : profiles of organic farmers and ranchers across the state / by Pamela Walker ; photographs by Linda Walsh.
p. cm. — (Texas A&M University agriculture series ; no. 11)
Includes index.
ISBN-13: 978-1-60344-107-0 (flexbound with flaps : alk. paper)
ISBN-10: 1-60344-107-7 (flexbound with flaps : alk. paper)
1. Organic farming—Texas. 2. Organic farmers—Texas—Biography. 3. Family farms—Texas. I. Walsh, Linda (Linda M.) II. Title. III. Series: Texas A&M University agriculture series ; no. 11.
S605.5.W33 2009
381'.4109764—dc22
2008042580

To the memory

of my beloved maternal grandparents,

Pa and Memmy,

also known as

Frank Leonard Whitis and Beulah Mae Ellis Whitis.

—Pamela Walker

Contents

Foreword

Allan Jones

While reading *Growing Good Things to Eat in Texas* on a Sunday afternoon in March, I couldn't help thinking about my grandparents' small dairy outside Comanche, Texas, in the 1950s. They, like most of the farm families described in this excellent contemporary account of organic farming and ranching in the Lone Star State, came to farming late in life. They endured many hardships in making the transition from managing a small grocery to milking a small herd of dairy cows, but they persisted, making a meager living but helping instill in their grandchildren a love of animals, the outdoors, and the smell of freshly baled peanut hay.

Pamela Walker takes a personal approach to describing the family businesses featured in this book. She weaves together biographical information, photographs, and the details of farming, aquaculture, ranching, dairying, food processing, soil and livestock management, and marketing techniques. The results are intimate stories of families struggling to realize their dreams in the face of economic and personal challenges and an agriculture and food marketing system designed by and for large-scale conventional (non-organic) agriculture. Included are the first certified organic citrus producer in the Valley; several vegetable and flower producers serving customers in Austin and Houston; a shrimp farmer in the Trans-Pecos; and poultry, beef cattle, and cheese producers in northeast Texas.

We are continually reminded that in families everywhere, elders pass on, children grow up and leave home, businesses evolve, and the difference between success and failure can be quite small. But these families, while pursuing organic agriculture and a more sustainable lifestyle, are themselves sustained by their love of the land, a strong work ethic, close family ties, and the belief that what they do for a living is worth more than the income it produces.

Yet several disturbing themes recur in the real-life experiences of these families. These include the cost of organic certification by government agencies, the failure of federal and state agriculture agencies and state universities to meet their needs for technical and regulatory assistance, and the inability of many families to afford health insurance or amenities taken for granted by most Texans. Instead, many of these families have found technical assistance from other organic farmers and ranchers, organic farming advocates, and a few county extension agents and businesses dedicated to sustainable agriculture. They have also received strong emotional support from their customers and churches, and they are sustained by the belief that they are doing the right things for their land, their animals, their families, and the people who consume their products.

In many ways these organic producers are struggling with the same challenges faced by most small-scale agricultural producers in the United States: government policies and programs designed to assist large-scale producers using conventional agricultural practices. These programs have produced abundant agricultural commodities at low prices, but they have largely severed the relationship between the families who consume and those who produce our food. And they have created a food system dependent on chemical fertilizers, synthetic pesticides, and animal health products. These small-scale organic producers represent a reaction against conventional agriculture. They are striving to reestablish relationships with families who consume their foods, provide them with more natural and health-promoting products, manage their land more sustainably, and treat their livestock more ethically.

This is not a book filled with statistics or details of organic agricultural techniques. It is not a compilation of how to be a successful organic farmer or rancher. It does, however, tell the stories of men, women, and children who make up the organic agriculture movement in Texas. All of us interested in Texas agriculture and rural life will gain from the experiences of the dedicated families whose lives are profiled in this book.

Acknowledgments

This book grows from my lifelong love of farmers, my maternal grandparents in particular, and my love of food fresh from fields and gardens. The single greatest pleasure in creating this book was visiting farmers and ranchers throughout Texas, walking and talking with them along rows of vegetables, through orchards, across pastures, and on pond levees. I am grateful to everyone who spared precious time for these visits. Their number includes not only those whose stories are in this book but also the others I wish I could additionally have included: Frank and Pamela Arnosky; Jackie Bass; Robin and Peter Bowman; Carey Burkett and Steve Kraemer; Charlie and Elizabeth Gearhart; Bay Laxson; Gayla Lyons and Edgar Chavez; Tim and Melody McCullough; Tim Miller; Jacky and Cindy Morrison; Tony and Suzanne Piccola; Dan and Mary Rohrer; and Gary and Sarah Rowland.

I am also grateful for conversations with many people who, in various professional roles, serve the advancement of organic and sustainable agriculture in Texas and the nation and provided me with valuable insights and information: Douglas Constance, Associate Professor of Sociology, Sam Houston State University, Huntsville, Texas; Susan DeMarco and Jim Hightower, Hightower & Associates, Austin, Texas; Richard de los Santos, State Coordinator for Horticulture, Produce and Forestry Marketing, Texas Department of Agriculture, Austin, Texas; Monty Dozier, Regional Program Director, Texas AgriLife Extension Service, Texas A&M University, College Station, Texas; Wylie Harris, rancher in St. Jo, Texas, and Contract Communications Specialist at the Kerr Center for Sustainable Agriculture, Poteau, Oklahoma; Art Ivey, pecan farmer, Rio Bravo Farms, Tornillo, Texas; Jonathon Landeck, Deputy Executive Director, Organic Farming Research Foundation, Santa Cruz, California; Leslie McKinnon, Coordinator for Organic Certification, Texas Department of Agriculture, Austin, Texas; Dan Nuckols, Associate Professor of Economics and Business Administration, Austin College, Sherman, Texas; Max Pons and Lisa Williams, biologists at the Nature Conservancy's Lennox Foundation Southmost Preserve, Cameron County, Texas; Robert Sandner, independent organic agriculture consultant, Uvalde, Texas; Suzanne Santos, Program Director, Farm Marketing, Sustainable Food Center, Austin, Texas; Richard Sechrist, rancher, Fredericksburg, Texas; Barbara Storz, AgriLife Extension Agent for Horticulture, Hidalgo County, Edinburg, Texas; and Granvil Treece, Aquaculture Specialist, Sea Grant College Program, Texas A&M University, College Station, Texas.

Shannon Davies, Louise Lindsey Merrick Editor for the Natural Environment at Texas A&M University Press, played a singularly instrumental role in making this book a reality. Though skeptical of my initial proposal, she generously remained open-minded and, as the book evolved, became an enthusiastic supporter and made its publication possible. For this, I am grateful. For guiding the book through editing, design, and production,

I am indebted to Jennifer Ann Hobson, project editor, Texas A&M University Press, and to John Thomas for copyediting.

Friends and members of my family also helped me enormously, and I am grateful to each of them. My two sons and daughters-in-law, Ben Johnson, Andrew Johnson, Michelle Nickerson, and Catherine Korda, read parts of the manuscript in the early stages and raised useful questions. So did friends Paul Harcombe, Mary Kaye Jennings, Colleen Morimoto, Susan Murray, and Irene Pendergrast. Nathaniel Norton lent me his guestroom during a couple of trips to the Rio Grande Valley. And Irene Pendergrast and David Morris, to help defray the cost of reproducing color photographs, aided me in researching possible funding sources.

Walter Isle, my husband, helped me most of all, and to him I owe the greatest debt. Without his unflagging intellectual, emotional, and financial support, I could not even have begun the travel and other work that went into this book.

—Pamela Walker

I give thanks to the following people:

Pamela Walker, for inviting me to participate in this labor of love, for her hard work and intellect, for her diligent attention to detail and organizational skills, and for her extensive research and rigorous selection standards.

The farmers and their families, who shared so generously and freely of their farming practices and knowledge, their personal histories and family lives, their time, and their willingness to be photographed for this book.

Shannon Davies, Louise Lindsey Merrick Editor for the Natural Environment at Texas A&M University Press, who made the idea of this book a reality.

Peter Brown, for years of guidance while gently critiquing my work, for always offering possibilities, taking time to listen and to offer practical solutions, and supporting and promoting me and my work at all stages of my professional development.

Jerome Crowder and Paul Zeigler, for generously sharing their expertise, equipment, practical solutions, and time.

Geoff Winningham, for teaching me how to read color and to bring it out on paper and for continuing to offer advice and share his knowledge.

Jean Caslin and Diane Gregory, for an early "read" on my photographs and for believing in the work from the beginning.

Madeline Yale, Rachel Hewlett, and the Houston Center for Photography, for training and the digital lab.

Patricia Eifle and Jim Belli, for generously opening their home to the Cream of Rice photography group. It has been an important center of sharing and learning.

4W members Debra Boylan, Farrah Braniff, Jim Dilger, Linda Gilbert, Geraldine Gill, Ruth Heikkla, Charlotte Randolph, RoyceAnn Sline, David Vaughn, Rod Waters, David Williams, and Dorothy Wong, and Cream of Rice members Julie Alexander, Mark Bagge,

Deborah Bay, Marty Carden, Peter Chok, Kelly Dempster, Jim Dunlap, Jim Falick, Abbie Flynn, Tom Foster, Val Glitsch, Naveen Jaggi, Sharon Jones, Len Kowitz, Craig Mabrito, David McClain, Luis Mesa, Yousef Panahpour, Adrienne Patton, Kristy Peet, Alan Schlesinger, Lou Smith, Bill Snypes, Suzanne Street, David Veselka, Bill Walterman, and Delise Ward. I have learned from each of you.

Stuart Allen, Scott Martin of Onsight, and Mary Ann Jacob of Texas A&M Press, for technical support, advice, and encouragement.

Pam Barton, Susan Speight, and Nell Wegmann, for teaching me never to say no to an adventure, including this one.

Newton Hightower, Sandra Jonas Desguin, Jenn Rudgers, Sandra Wegmann, and Ken Whitney, for listening.

Steve Barry, Richard Brooks, Robin Brooks, Dan Kane, Sam Kane, Sheila Kane, Bill Kelley, Joy McCormack, and David Walsh, for always being there.

Anita Walsh, my farmer and photographer sister, my beacon of excellence, my most honest critic, and my inspiration.

Terrell Dixon, whose love, encouragement, humor, enthusiasm, and support made it possible for me to take on this project in the first place and to see it to completion.

—Linda Walsh

Growing Good Things to Eat in Texas

The cure for digging in the dirt is an idea;

the cure for any idea is more ideas;

the cure for all ideas is digging in the dirt.

—Kenneth Burke

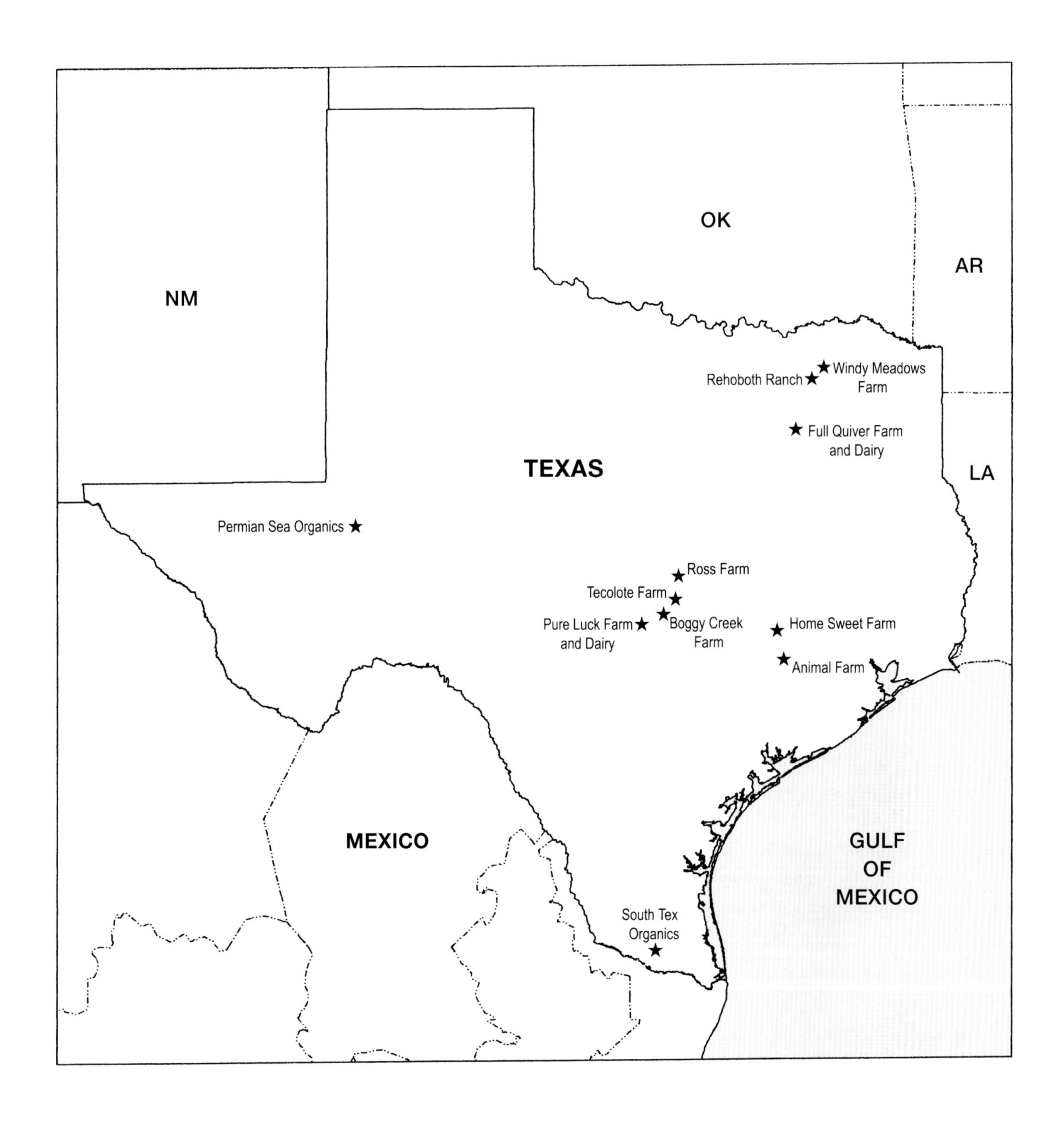

NM
OK
AR
TEXAS
LA
Windy Meadows Farm
Rehoboth Ranch
Full Quiver Farm and Dairy
Permian Sea Organics
Ross Farm
Tecolote Farm
Pure Luck Farm and Dairy
Boggy Creek Farm
Home Sweet Farm
Animal Farm
MEXICO
GULF OF MEXICO
South Tex Organics

Introduction

This book of essays and photographs features eleven farming and ranching families in Texas. The essays are short biographical stories describing the food these farmers and ranchers produce, how they produce it and market it, and how they got into doing the work they do. But the stories are also about you and me and everyone who eats. When we purchase food produced by people such as those in this book—real food, food grown organically, in ways that nourish our bodies as well as the soil, and, if we're especially lucky, grown close to where we live—we aren't merely eating well. We're helping to preserve and extend vibrant farmland and independent farming as a way of life. And we're helping to sustain diverse social and economic communities that arise from the growing and eating of real food.

How did I come to find the farmers and ranchers featured in this book, and on what basis did I select them? The answer is a farm-and-food story all its own, and one that begins about six years before I first envisioned this book. In late 1997, my husband and I purchased a small farm near Schulenburg, a hundred miles west of our home in Houston, and started dividing our time between the two places. My desire to live at least half the time on a farm stems directly from the fact that my grandparents were farmers, sharecroppers who grew cotton in Central and North Texas and finally southwestern Oklahoma. Until my mid-thirties, when they died, I visited them often, for weeks at a time during my childhood, and from them learned to love vegetable gardening and eating food fresh from the ground. No matter how tiny my inner-city yards have been, I have always grown vegetables organically and maintained a compost heap. Acquiring a farm allowed me to garden and compost on a much larger scale than ever before. At last, I had space enough for an asparagus patch, for long rows of black-eyed peas, for twelve tomato plants or twenty, if I wanted, instead of two or three, plus space enough for cantaloupes and pumpkins and their long, winding vines.

Enjoying a greater and more constant variety of fresh vegetables than ever before made me wish that farmers' markets were common throughout Texas, as in some other states, so that local produce could be available to everyone, not just gardeners and farmers, and be an ordinary delight instead of an extraordinary one. Wouldn't it be wonderful, I thought, for all of us to have farmers' markets to weave into the fabric of our weekly routines. Well, I pretty soon figured out that wishing for farmers' markets where I wanted them was not going to make them materialize, and by early 1999 I was researching market organization and development. I did a lot of reading, and I visited as many farmers' markets as I could, both in Texas, especially the Austin area, and out of Texas, during vacations and other travel. I also joined the Texas Organic Farmers and Gardeners Association (TOFGA) and Holistic Management International–Texas (HMIT) and began attending their conferences and other events. The upshot? I was fortunate to be instrumental in the opening of a small farmers' market in Schulenburg in 2000, and from 2002 through 2006 I led in the organization and initial development of the Bayou City Farmers' Market (a project of Urban Harvest, a Houston community gardening nonprofit), which opened in September 2004.

In the course of this market work, I found myself increasingly thinking that the stories of the farmers and ranchers I met at markets and conferences, and the significance of their achievements, were too compelling and instructive not to be publicized. I wanted more people in Texas to know where to buy Texas produce and to become familiar with the fascinating people producing it. I wanted people thinking about entering organic agriculture themselves to see that, not only can it be commercially successful here, it already is—and in many cases has been for quite some time. And we would all benefit from its expansion. A book of biographical essays based on personal interviews, and complemented by photographs, was the way to go, I decided early in 2004, and I asked Houston photographer Linda Walsh, a longtime acquaintance, to participate in the venture.

To demonstrate the viability of commercial organic agriculture for family-based producers, whether certified organic or not, I made it my main criterion that those selected for interviews must make all or a substantial part of their living in organic agriculture and must have been doing so for at least five years. To identify as many potential interviewees as possible and not limit my consideration to those I already knew or knew of, I did several things. I asked TOFGA and HMIT members for the names and contact information of producers who fit my scope, and I also asked for producer lists from the organic program staff of the Texas Department of Agriculture (TDA). This process produced an initial list of about 150 food producers.

My next step, in the late spring of 2004, was to write a one-page letter describing the project and devise a short, simple questionnaire, which I either mailed (with a return envelope) or e-mailed to those on my list. I asked seven questions: (1) What do you produce, and on what size acreage? (2) How long have you been a commercial producer? (3) About what proportion of your income is from farming or ranching? (4) If you are also employed away from your farm or ranch, what work is this? (5) How do you market your products? (6) Are you certified organic? (7) If you know other producers who fit the scope of my book, would you please list their names and contact information?

Nearly half of those receiving my letter and questionnaire replied. Compiling their information with information I already had produced a list of ninety farmers and ranchers to cull from. In addition to my main criterion of established commercial success in organic agriculture, my choices were guided by a desire to end up with a range of crops and state regions represented in the book. Unfortunately, I received no responses from family-based organic producers in the Panhandle, but I did get replies from people in all other regions, and from these I selected twenty-five farmers and ranchers to ask to visit. Twenty-four generously agreed, and Linda and I visited them between August 2004 and December 2007, some more than once, for a total of more than thirty visits.

In each case, we spent hours walking around the farm or ranch, and Linda took photographs as I talked with the farmers and ranchers in detail about their work and lives. "What do you do? And how did you get here?"—these were my basic questions. Equipped with a small cassette recorder, I listened and asked many other questions. Between visits, I transcribed all of these conversations, and as the long process of visiting and transcribing went on, I gradually figured out which farmers and ranchers to include. Which stories, brought together, I kept asking myself, best served my main goals for this project?

I regret that I could not include everyone, but, alas, books can only be so long, and I had to make some hard choices. The result is a book of ten biographical essays and accompanying photographs, nine of the essays focusing on a single family and one of them (Rehoboth Ranch and Windy Meadows Farm) on two families who are friends and business associates. I have written these essays to be read in any order and grouped them into three product categories. The first, "Fruit and Vegetables," includes five essays. "Shrimp and Meat" is covered in three essays. The third, "Dairy and Cheese," has two essays. Within each category, the essays are arranged more or less chronologically according to the time of the farmers' and ranchers' entry into substantial commercial production.

The stories in this book represent not so much a looking back, however, as a looking forward. Organic agriculture, though miniscule compared with conventional agriculture (only 0.5 percent of total U.S. cropland and rangeland, or pasture), is growing faster than any other sector of American agriculture. This growth includes both large-scale organic operations with international wholesale markets and small organic operations with direct retail venues such as farmers' markets and community-supported agriculture arrangements (CSAs), subscription programs in which customers pay a farmer for produce in advance of a season and underwrite some of the production costs. For the past decade, the number of certified organic producers has increased nationally at a rate of about 20 percent a year, to about 8,500 operations on a total of four million acres, including about 192 operations on a total of approximately 328,500 acres in Texas.

Although the growth in the number of producers who use organic practices but do not certify their land cannot be exactly tracked, the increase in the number of farmers' markets, a prime venue for most small, uncertified producers, indicates a similarly significant expansion. Nationally, the number of farmers' markets has grown about 18 percent since 2004 (from about 3,700 to 4,400). During the same period in Texas, the number of markets certified by the TDA has also grown by about 18 percent (from 94 to 111). Because TDA certification is optional, however, figures for TDA-certified markets do not account for any uncertified markets, and so the total number in any given year in Texas is always somewhat higher. CSAs, though not officially tracked by government agencies, also are increasing. The respected nonprofit organization Local Harvest, founded in Santz Cruz, California, in 1998 to provide a national online directory of small farms and market venues, estimates that there are now more than a thousand CSAs in the United States, up from about fifty in 1990. Thirty-seven CSAs are currently listed for Texas.

The expansion of organic agriculture is being driven not from the top down but from the bottom up, by growing numbers of people—like the farmers and ranchers in this book and those of us buying their products—who are making the connection between the healthfulness of what we eat and the health of the soil that nourishes it and choosing organic products. Organic food purchases account for 3.5 percent of the U.S. retail food market today and are expected to reach 10 percent by 2012. At the "top," however, resources devoted to organic agriculture are scant, lagging way behind customer demand, and many organic education and advocacy organizations are consequently pushing for changes that would give organics a fair share of public monies. The Organic Farming Research Foundation (based in California but national in programming scope), for example, is pushing for the U.S. Department of Agriculture (USDA) to keep its budget for research and education on organic agriculture aligned with the percentage of retail food purchases annually going to organic food. Were this done today, the USDA would be giving organic research not 1.2 percent of its budget (about $26–$28 million) but 3.5 percent (closer to $80 million). The European Union, by contrast, currently spends about $100 million annually for organic research.

TDA devotes an even smaller proportion of its budget to organic agriculture, funding a single small office to administer its USDA-accredited organic certification program. The program staff includes only three people: a coordinator and two administrative assistants. For fiscal year 2007, only 0.7096% of TDA's general revenues, or $0.71 for every $100 in its budget, went to the organic certification program. Although most certified organic producers in Texas are certified by TDA, some choose accredited agencies outside of Texas that have more funding and staff and can process certification applications and renewals faster. The TDA organic program, in a new effort to mitigate the slow service resulting from underfunding and understaffing, has contracted the part-time assistance of a staff person from the Independent Organic Inspectors Association, a nonprofit association providing support services for certification agencies. And, in early 2008, TDA established an Organic Agricultural Industry Advisory Board of thirteen people, including four organic producers, to guide TDA on such matters as improving organic marketing functions, developing official TDA comments to the USDA whenever changes are proposed to the national organic standards, and recommending ways to promote the Texas organic industry.

Another publicly funded agency that makes only a small amount of money available for organic agriculture is the cooperative extension service, now called the Texas AgriLife Extension Service. Funded by federal, state, and county monies and administered through Texas A&M University for the purpose of providing public education programs on agriculture, AgriLife services, like all extension services, are by law based on the agricultural research conducted at land-grant universities and consequently overwhelmingly promote conventional agricultural production and marketing methods. However, through a USDA program called Sustainable Agriculture Research and Education (SARE), AgriLife receives $40,000 of grant money annually for programs to help extension agents learn more about sustainable agriculture. Half of this money is for a half-time staff person to organize program events, and half is for all other program expenses. As defined by SARE, sustainable agriculture includes organic methods based on the national standards as a subcategory but also includes "low-input" chemically based conventional methods. So the amount of these funds actually supporting organic agriculture is infinitesimal.

Yet even without official funding through AgriLife, some remarkable organic agriculture programs arise from the extension service, as individual agents form partnerships outside the agency and get outside funding. An outstanding example is the Grow'n Growers Program in the Rio Grande Valley, organized and run by Barbara Storz, horticulture agent in the Hidalgo County extension service, in Edinburg, and involving multiple private and public funders. Grow'n Growers is a nine-month course of twice-weekly classes on organic vegetable production and marketing, food safety and handling, and small-business development. Storz teaches the course, assisted by other AgriLife agents and specialists, master gardeners, local growers, and local TDA staff. Participants are low-income Hispanic people, mainly women, who to be selected for the program must commit themselves to attending all the classes, growing their own vegetables, and joining a marketing co-op. Called Familias Productores del Valle, or Family Producers of the Valley, the co-op is a project of Heifer International, a nonprofit organization that fights hunger and fosters economic independence by assisting people all over the world to raise their own food. For Storz's Grow'n Growers class, Heifer provides the money for seeds, tools, and materials for raised garden beds in each participant's yard. At the end of the course, graduates become vendors at an all-organic market Storz led in organizing, the San Juan Farmers' Market, in the small city of San Juan's North San Juan Park. The park is also where Storz conducts the Grow'n Growers classes, in a semi-open outdoor classroom surrounded by raised demonstration beds, facilities she funded by partnering with the Hidalgo County Commissioners Court and the City of McAllen compost operation.

In addition, Storz has headed a sustainable agriculture task force since she became an agent, in 1998. Drawing about 150 people from all over the Valley, the group is made up of certified organic growers, growers in transition from conventional to organic production, commercial composters and soil analysts, and nonprofit community developers. They meet monthly for a lunchtime program of speakers and follow-up discussions. "All across the country, the Valley included, consumers are driving the demand for organic," Storz says, "and we extension agents have a lot of freedom in how we respond to our community's needs."

Standing apart from publicly funded agencies, the only entity in Texas wholly devoted to promoting organic agriculture and to nurturing a statewide community based on organic food production and consumption is the grassroots organization TOFGA, a small, all-volunteer operation with no paid staff. It was founded in late 1993 as the Texas Organic Growers Association (TOGA), five years after the 1988 establishment of TDA's organic certification program, one of many certification programs throughout the United States predating USDA's 2002 implementation of national organic standards and the "USDA Certified Organic" label. TDA's Certification program was created when Jim Hightower was commissioner of TDA and made organic production and marketing a priority. To complement TDA's services, several organic growers, including Dennis Holbrook, who is profiled in this book, established TOGA with the intention of developing it into an organic marketing cooperative. With so few organic growers actively involved, however, and these so far flung across the state, TOGA never functioned as a co-op.

It evolved instead into a nonprofit association dedicated to bringing together not just organic farmers and ranchers but everyone interested in supporting organic growing in Texas—farmers' market developers, chefs and caterers, retail grocers, food and agriculture journalists, community and school gardeners, commercial organic composters, individual consumers—and in late 2003 it changed its name to Texas Organic Farmers and Gardeners Association. Under the leadership of its immediate past president, Brad Stufflebeam, a farmer who is profiled in this book, and before him, Steve Bridges, a Bastrop-area commercial horticulturalist, TOFGA has grown and become increasingly effective. It now has five hundred dues-paying members and a monthly e-mail newsletter list of more than 1,500. TOFGA's website (www.tofga.org), created by Stufflebeam, lists events throughout the state and also includes a "Local Food Locator," a directory of family farms and ranches organized by regions of Texas.

TOFGA's annual Texas Conference on Organic Production Systems is the largest conference devoted exclusively to organic agriculture in the state (and one of the largest in the South) and features as speakers distinguished people from Texas and many other parts of the nation—farmers and ranchers, nutritionists, representatives of nonprofit organizations involved in agriculture research and food-and-farm policy reform, certification professionals, to cite just some. The 2008 conference attracted more than 350 attendees. At these events, people who likely would not otherwise meet each other debate, both on panels and informally, topics such as those discussed in this book: the strengths and weaknesses of the national organic standards; the benefits and drawbacks of certification for particular kinds of producers; the comparative advantages and disadvantages of wholesale marketing and direct retailing; the meaning of such terms as "local" and "sustainable"; the hindrances to producers and consumers posed by certain health ordinances and other bureaucratic regulations; labor issues, and so on. In addition to these conferences, because many more organic producers are needed to meet the growing demand for organic food, TOFGA, like Storz, works to grow more growers. It periodically hosts workshops for new and prospective organic farmers and ranchers, bringing expert producers together with novices for technical discussions about production and marketing.

TOFGA's statewide leadership in promoting organic farming and food is unique in Texas, and vital. The financial donations and active involvement of organic food producers and consumers are essential to its further development and to its becoming an even more effective organization.

Along with supporting TOFGA, how else do we keep this good-food movement, this real-food revolution, growing here in Texas? How can we personally help grow more growers like the ones featured in this book? One is by buying as much of our food as we can directly from farmers and ranchers themselves, at farmers' markets, at farm stands, and through CSAs. Producers who sell directly to the public net almost four times as much from each dollar of sales they do by selling wholesale (approximately $0.86 net profit per dollar compared with $0.23). That is why most small producers depend mainly or totally on direct retail markets; they can't make a living otherwise. It's fine to buy organic produce at retail grocers (and probably none of us will ever stop going to grocery stores), but with rare exceptions most organic food in grocery stores, including Whole Foods and Central Market, overwhelmingly comes from industrial-scale growers in California or from other countries, which means that we are only rarely supporting Texas farmers and helping to preserve or expand farmland here in Texas with those purchases.

What if you don't have farmers' markets or CSAs in your part of Texas? Well, organize some. Many CSAs arise because customers wanting organic food approach and work with a grower to get one going. Starting a farmers' market is generally tougher. Developing any retail business is long, hard work, and the process can take a few years before the market even opens, but it is well worth the effort. Excellent resources about farmers' market organization and development are available right here in Texas, especially from the Sustainable Food Center, in Austin, the nonprofit organization that operates the Austin Farmers' Market. But what if you can't find many, if any, farmers, in your area interested in a local market? Start with gardeners and keep talking to any farmers you identify, no matter how few, familiarizing them with established markets elsewhere as models of the economic opportunity available from this type of venue. Talk with your county extension agents, too, and explore partnerships like those Barbara Storz formed in Hidalgo County. And share this book with all those people, to illustrate how viable, and valuable, organic agriculture in Texas is.

In addition, we need to vote for Texas-grown organic food not just with our food dollars but also at the ballot box. The commissioner of TDA is an elected position. Let's make organic farming and food an issue during campaigns, and likewise with our state and national legislators. To keep posted on the many regulatory issues that arise and threaten the economic viability and therefore the existence of small farms and ranches, join a new, Austin-based national organization called the Farm and Ranch Freedom Alliance. Its executive director makes personal visits to legislators and regulators in both Austin and Washington and keeps the national membership posted on such things as farm bill provisions and the proposed national animal identification system, noting key legislators to write or call on behalf of preserving family farming and ranching.

What else? Think about growing at least some of your own food even if you do not want a full-fledged vegetable garden. Grow some herbs; plant some fruit trees. And all of us, gardeners or not, can make compost from table scraps and weeds and leaves and use it to feed whatever little spot of the good earth it is that we call home.

Resources for more information

Farm and Ranch Freedom Alliance
www.farmandranchfreedom.org
This website provides current information about legislative and regulatory issues affecting independent farmers and ranchers in Texas and the United States and posts notices of educational events.

Local Harvest
www.localharvest.org
This website lists farmers' markets, CSAs, family farms, restaurants, grocery stores, co-ops, and other sources of sustainably grown food for each state.

Texas Department of Agriculture
www.picktexas.com/farm_market/farmers_market2.htm
See this website for a city-by-city list of farmers' markets in Texas that are certified by TDA.

Texas Organic Farmers and Gardeners Association
www.tofga.org
This site provides information about organic agriculture in Texas, including educational events and names and locations of organic producers and others involved in organics.

Fruit and Vegetables

Grapefruit tree, South Tex Organics.

South Tex Organics

Dennis Holbrook's Journey from Pharmaceutical to Organic Farming

It is late January in Mission, Texas. The Rio Grande Valley citrus harvest has reached its peak and is moving along full force. In the groves, packing shed, and offices of Dennis and Lynda Holbrook's South Tex Organics, everybody is busy. Dennis's office, where most of our conversation takes place, is no retreat from all that is going on but the main hub. It is filled with a dark wood desk and cabinets and situated directly across the loading area from the packing shed. Dennis, boots on and field jacket always in reach, goes out from time to time to see that fruit coming from the groves is carefully unloaded for washing, sorting, and labeling, then packed into boxes and placed onto wholesalers' trucks according to specifications. And even though Dennis has a full-time administrative assistant at work in an office adjoining his, he takes some phone calls himself, sometimes pulling documents from neat stacks on his desk for reference. South Tex Organics is large and multifaceted, employing more than forty people, and Dennis, a fifth-generation farmer, keeps his hands in every part of the business.

Ask him what he enjoys doing most, and he doesn't have to search for an answer. "Being involved in the whole evolution of a tree," he replies. "I love it. Preparing the land and nurturing the soil, planting the tree and helping it grow, watching it bloom and bear, and then harvesting the fruit." And he harvests lots of it, three hundred certified organic acres of grapefruit, orange, and lemon trees, from November into March of each year. Varieties include Texas Rio Star and Henderson Honeygold grapefruit; Marrs, Hamlin, and Valencia oranges; and Meyer lemons. In addition, from March into June, on another hundred certified organic acres, he raises red and yellow onions, watermelons, and honeydews. Using a system of large, underground pipes with surface valves, he irrigates by flooding designated irrigation rows with water, the standard method in the Valley. He considered a drip system but decided that, on such a large scale, he could do a better job with flood irrigation, conserving water by improving retention and reducing irrigation frequency.

The soil is basically a sandy loam, and the care Dennis gives it and the trees is everywhere apparent, in the size and sheen of the leaves, in the luster and abundance of the fruit, and in the bands of alfalfa, sorghum, legumes, and native vegetation—the green manure crops periodically tilled under to decompose and feed the dirt—in and around the groves and vegetable fields. But probably the most eye-catching signs of how much care the soil gets are the big compost piles across from the packing shed and offices. Pale yellow heaps of ground corn and cobs mixed with dry molasses occupy one end of the area. Warm to the touch and yeasty smelling, this coarse meal is residual material from a Mexican fruit fly project at the nearby USDA research station. Opposite the yellow heaps are dark brown mounds of tree mulch from the McAllen utility maintenance district, and in the broad space between the mulch and the meal is a bright low hill of culled citrus, a jumble of oranges, grapefruits, and lemons that have been damaged or are otherwise unfit for consumption.

"We apply compost to the trees shortly before they bloom," Dennis explains, "from the end of February through March. As the trees come out of dormancy, they're exploding with growth. They're putting out a lot of feeder roots, and they need plenty of nutrients. We have a front end loader on a backhoe and what we do is scoop and mix and push and roll the piles together, over and over and over, till it's mixed up. And then we'll scoop it up and put it in our compost spreader and away we go."

Dennis, a medium-built man with blue eyes and a receding hairline, was born into a Mormon family in Utah, far from citrus country, but his life has revolved around citrus groves and row cropping for longer than he can remember. He was eighteen months old when in 1955 his family moved to the lower Rio Grande Valley—"I got here as soon as I could," he laughs—and his father, who in Utah had raised cattle and hay, started a grove management business, planting and maintaining groves for landowners who lived too far away to do it themselves. He also purchased a farm for row crops and grew a wide variety of vegetables and melons on it. In all his farming, he followed conventional methods and used synthetic chemical fertilizers, pesticides, and herbicides.

Growing up, Dennis, the younger of two sons, took more interest in farming than his brother did, and from an early age he worked often in the groves and fields. He did everything his father and the farm workers did. He tilled, planted, irrigated, operated spray rigs, and harvested. He learned to weld and fix implements with moving parts and to repair the diesel engines of trucks and tractors. He enjoyed the work and was good at it, good enough that at the age of sixteen he ran the entire business for a month, including payroll, while his parents traveled in Europe. This experience confirmed for Dennis that he really did want to do what he had always figured on doing, go to college and major in some aspect of agriculture and then return and work with his father.

Hardly three years later, however, toward the end of Dennis's freshman year at what was then Texas A&I, in Kingsville, his father surprised him with news that he had decided to sell his grove management business. Additional businesses in land leveling and real estate were claiming more and more of his father's time, and something needed to go. A buyer had already approached him, but he wanted Dennis to have the first option. "I was nineteen years old," Dennis emphasizes, as if still a little incredulous that he was so young and suddenly required to make a now-or-never economic decision. "But I'm a very commonsensical individual," Dennis explains, "and I analyze things. I think okay, here are all the benefits, and here are all the negative parts. I mean, no one has a crystal ball, but it made sense to me to go back and work and start purchasing the business."

Back in the Valley, Dennis continued with college for a while, transferring to Pan American University in Edinburg, but the demands of farming, followed soon by the responsibilities of marrying and starting what eventually became a family of four children, quickly won out over those of school. Dennis and Lynda met at Pan American, where she was studying social work. Born and raised

Lynda and Dennis Holbrook in the South Tex Organics office.

in Edinburg, in the midst of crops of all kinds, Lynda, a small woman with dark brown hair and eyes, paid little attention to farming before getting to know Dennis. Her father owned a car dealership, and her mother managed the office of another businessman. "I saw vegetable fields and citrus groves every day of my life," she recounts, "but I never gave it any thought. I just bought oranges and grapefruit and everything else in the grocery store. Until I married Dennis, I had no idea how groves were taken care of, and I found it really fascinating."

Dennis gradually shifted the business away from managing others' groves to cultivating his own and raising row crops on open land. For the first decade, he farmed conventionally, as his father had. But, by 1980, dwindling profits and a growing concern about the hazards of agricultural chemicals to human and environmental health had pretty well convinced him that switching to organic methods was the most sensible way to farm.

He had long noticed that the quality of the soil was

Inside the South Tex Organics packing shed.

getting worse and worse, compared with how it had been when he started working with his father in the sixties. Back then, the Holbrooks had used less herbicide for weed control than in subsequent years and had relied more on tilling weeds under, which helped check their spread and also, as they decomposed, became humus that fed the soil and nourished plants. These tillage practices also improved water retention and promoted deep root growth for citrus trees and row crops, which kept the number of irrigations down to about five a year. But as the Holbrooks increased their use of synthetic chemical herbicides, pesticides, and fertilizer, they depleted the soil of humus and water. This created a hardpan surface in some of the groves, soft, shifting sands in others, and, throughout all the groves, poor water absorption and trees with shallow root systems. By the late seventies, the Holbrooks irrigated as many as ten times a year.

Soil tests only substantiated the nutrient depletion that Dennis could see with his own eyes and feel in his hands. Nowhere could he find the rich, moist crumbliness of a healthy soil. The humus content in some of his groves was as low as 0.22% and barely higher in others. Wondering what the level of organic matter had been before chemical farming became the norm, Dennis consulted historical records at a local soil science laboratory and learned that in the late forties and early fifties a humus content of 2.5% was common in soils throughout the Valley.

"I realized that we were working contrary to Mother Nature, not in conjunction with her," Dennis says. "We were spending more and more money on chemicals and irrigation, making less and less of a profit, and doing a lot of damage in the process. Growing with chemical weed control and chemical everything else, we had actually created almost a sterile medium where we were having to input everything synthetically into a tree to produce a crop. We were basically making junkies out of the trees. They weren't healthy or productive."

In the course of seeing what chemicals did to the soil and plants, Dennis couldn't help but wonder what they were doing to the people who handled them, himself included. Having operated spray rigs since he was a teenager, he decided to participate in a study conducted by researchers from Texas Tech that involved analyses

Dennis Holbrook checks packing-shed machinery.

of farm workers' blood, urine, and hair for agricultural chemical content. It turned out that Dennis had a high content of various chemical residues, higher than any of his workers. It was disturbing but not surprising, just one more part of the predicament. "Nothing I could do about what had already happened, of course," he reflects, "but there was no getting around the need for change. I knew I had to switch from pharmaceutical farming to organic. The big question was how. Just how did you get off those chemical merry-go-rounds?"

A severe freeze in 1983 provided the solution, destroying so many groves that, to stay in the business, Dennis more or less had to start over. "It was the perfect opportunity to go cold turkey," he says. "It was either go organic or give up farming." So he pruned his trees to the ground and bulldozed and burned them up. "As a kid I always thought freezes and then the bulldozing and the bonfires were pretty neat," he remembers. "But in '83 I was no kid, and the business was mine and Lynda's and we had a son and a daughter and another baby on the way. It was pretty scary." Fortunately, they had some crop insurance, so the economic loss wasn't total. Dennis began replanting and also busied himself learning as much as he could about organic methods. In 1984, with sixty acres of newly planted groves, he and Lynda established South Tex Organics, the first organic citrus operation in the Valley. They received certification in 1988, when TDA began implementing its certification program, and in 1989 Dennis was appointed to the first TDA Organic Certification Review and Standards Advisory Committee. He has gone on to serve at the national level as well, appointed in 2002 by Ann Veneman, then secretary of the USDA, to the National Organic Standards Board.

The 1983 freeze also gave Dennis an economic opportunity to build his own packing shed, the first, and still the only, certified organic packing facility in the Valley. "After the freeze, a number of packing sheds didn't reopen, because the production wasn't there," Dennis explains. "Prior to the freeze, there were approximately seventy thousand acres down here in citrus, and only about thirty-five thousand after it. So we lost 50 percent of the industry, and there was all of a sudden a glut of equipment and not any buyers. So I was able to go out and buy key pieces of equipment for relatively little money. And I'm an individual who can go and watch a piece of machinery work for a while and see the whole operation and figure out how to change the design and components for what I need. So I would go study some of the sheds and then come over here to our place and design and build my machinery, and I put our shed together for probably less than $25,000."

Dennis and Lynda, as pioneers of organics in their area, have learned, both during the transition to organics and since, that organic farming and just the term "organic" itself can elicit ridicule and other sorts of dismissal. "When I went to the extension service for information and advice in the early eighties," Dennis remembers, "as soon as the word 'organic' came out of my mouth, I was looking at the door. They told me they didn't have anything."

"And Dennis didn't particularly care what it was called," Lynda points out. "He was pursuing alternative growing methods, a healthier way to do things, and it happened to be labeled 'organic,' and so we were labeled 'organic' and ostracized by some people. Even now, at annual meetings of the Texas Produce Association, which is mainly a Valley group, the after-dinner entertainment always includes some satire of organics, and Dennis is the only organic farmer there. It's like we've gone to the far side, even though before World War II that's how people farmed."

"Yeah," Dennis agrees, "and back in '83 what I realized I had to say to the extension people was look, you know, what I'm really asking you is to go back to the records for the forties and early fifties and tell me what the recommended growing methods were then. And that got me somewhere, got us on the path"—that and discovering Malcolm Beck, founder of Garden-Ville, a compost company based in San Antonio, and one of Texas's foremost practitioners and proponents of organic growing. "I called up to San Antonio and got Malcolm and told him what I was trying to do and he said come on up," Dennis recalls, "and I went up there and spent a day following him around like a little puppy dog, asking questions, and he was just a fountain of knowledge and information, especially about composting and how to attract and use beneficial insects."

TDA provided help, too, especially with marketing. "Jim Hightower [commissioner of agriculture at that time] was getting organics going in Texas, developing standards and so forth, and his marketing department was almost at our beck and call," Dennis says. "TDA priorities are different now, with little support for organics, but back then if you were organic and needed help marketing, you called up TDA and there was somebody to help you immediately." To complement TDA's marketing services, Dennis helped to found the Texas Organic Growers Association in 1993, which was intended to be a marketing cooperative. This intention was never realized, but the organization (now TOFGA) did help actual and potential organic growers identify and communicate with each other about marketing opportunities.

Even so, the Holbrooks found that establishing reliable markets for an increasing volume of citrus and vegetables was difficult. In 1986, their first year of organic citrus production, they sold everything to Whole Foods, which at the time was only one store in Austin, but no trucking service was available to haul the produce up from the Valley. "I happened to find a guy who was out of work and had a stock trailer," Dennis says, "and I said, 'Hey, if you wash that thing out and clean it up, I'll hire you to haul my stuff to Austin.' So that's what we did."

"The next year, we moved up in the world,' Lynda continues, laughing. "We hired a friend of ours with a meat truck!"

In those early years, Dennis also pursued venues even closer to home. He took boxes of citrus fruit around to groceries and health food stores in the Valley, hoping to set up wholesale accounts throughout the area. "But I couldn't get anybody interested," he says.

With his production increasing, he soon realized he needed to break into the large, better established California organic markets, though he was not optimistic. "I'm just one little guy over here in Texas, right?" he recalls thinking. "And California's where the organic thing is at, so it's this David and Goliath type thing, you know?" Nevertheless, he put together sample boxes of oranges and grapefruit with a cover letter and shipped them to eight different wholesale distributors in California and the Pacific Northwest. His first follow-up phone call was exactly what he had feared. "The buyer just totally blew me off," Dennis says. "He told me, 'We don't import into California. We export California.'" Three days later, however, the same person called Dennis back. He and others in his business had by then gotten around to opening the box of fruit from South Tex Organics. "'How soon can you start getting your product out here?' the guy asked me. They'd seen it, cut it, eaten it, and they thought it was great!" A giant had been won over, after all, and Dennis beams, savoring the moment even now. "So we started shipping truckloads to California, and from California some of our product started going to Illinois, Pennsylvania, and New York."

California wholesalers have been an important market ever since, but they no longer handle the majority of the produce from South Tex Organics, largely because Texas citrus is subject to quarantine during outbreaks

Workers cull fruit as it enters the South Tex Organics packing shed on a conveyer belt.

of the Mexican fruit fly, which means bans on shipping fruit to other citrus-producing states. "As my volume kept growing," Dennis recounts, "I told myself, 'You know what? You're really not being very smart, very business-like, because if you're shipping 50 percent of your volume to a place you could lose anytime there's a quarantine, then what are you gonna do with all that additional fruit?' So we decided we'd pursue other markets, build stronger markets in other areas where we could actually move more volume." Dennis now works with about forty wholesalers, and his produce is distributed not just throughout the United States but also in Canada, Europe, and Japan. "We're fortunate," Dennis says. "We can't even fill the demand that we've got. And that's one of the reasons we've expanded over the years."

Although wholesale marketing does not allow for a direct personal connection between the farmer and the people eating his or her produce in the way farmers' markets and other retail venues do, Dennis fosters the personal dimension through custom service and loyalty to his long-standing buyers. "One of the buyers from Kroger has been trying to get me to sell to them all season long, and I haven't been able to do that because I feel a commitment to the people who brought me here, the people who've actually enabled me to be in business and grow my business," he explains. "And to be perfectly honest with you, the produce business is probably one of the most disloyal businesses there is because it's all price driven. Most people say, 'Hey, if I can buy it over there for a quarter, I'm going over there, you know, cheaper is where I'm going.' But I appreciate the fact that we've had very good loyalty from the people who buy from us. But then we do our very, very best to make sure that we're a service-oriented company, you know, that we go the extra mile. I mean, if they call me at the eleventh hour and say they need to increase their order, we do all that we can to do that even though sometimes it's a real pain for us. We've had nights when, to get orders out, we've packed till two o'clock in the morning. Most packing sheds won't do that."

Compost pile at South Tex Organics.

Lynda points out that satisfying customers includes accommodating their preferences for certain sizes of fruit. "We don't go out and harvest our groves and have the fruit already packed in the boxes and say this is what we have to sell you," she elaborates. "We go out and harvest fresh. We try to harvest, pack, and ship all within the same day, within twenty-four hours. So our customers are getting the size they want, as fresh as we can get it."

"We've had times when we've needed another two bins of a certain size of fruit," Dennis adds, "and we're out there picking off the headlights of trucks at night, you know, trying to get what a customer wants, and nobody in this industry will do that."

Valuing this personal dimension is largely what prompted Lynda to develop mail order and Internet sales of citrus fruit directly to individuals, their sole retail venue. "It started as a kind of service to people, a favor," she explains. "Back in the late eighties and early nineties, I started getting calls from people telling me they couldn't get organic citrus fruit in their grocery stores and asking us if we could mail them some. The first couple of calls I got were really touching to me. I got a couple of phone calls back to back almost from two people who had just left M. D. Anderson after cancer treatment, and their doctors were recommending an organic diet."

Calls like these kept coming, and so did other sorts of special requests. Lynda and Dennis play competitive tennis, and one year the organizers of an annual Valley tournament asked them to donate gift boxes to all the tennis players. Among the recipients was a doctor who liked the fruit so much he immediately asked to buy a hundred boxes, enough for everyone in his Harlingen, Weslaco, and McAllen offices. "So it's just kind of evolved from what people ask for," Lynda says. "And with the Internet, it has really blossomed. The demand keeps growing, and I'm kind of like, what have we done here?"

Most of these individual sales, many of them holiday gifts, occur from mid-November into January, when more than a hundred boxes a day go out. Increasingly popular and successful as this aspect of the business is, however, it takes more and more of Lynda's time and that of Dennis's administrative assistant, and the income it brings in is small compared with wholesale accounts. Lynda and Dennis often wonder if it is worth continuing. Since it is Lynda's project, the decision is mainly hers and, whatever ambivalence she feels, she has not been willing to give it up. "I appreciate what we have. I really do," she says, "and I like providing our produce to individuals who appreciate it, too. I enjoy these personal connections."

As South Tex Organics moves toward its twenty-fifth year, it is among the larger organic citrus operations in the nation and is thriving more than ever. Dennis considers it a sustainable business not only because they continuously nourish the soil but also because they grow and sell on a large scale, and, beginning in the late eighties, they got into land development as a kind of insurance to mitigate the economic risks inherent in farming. In contrast with advocates of small-scale farming, including other farmers profiled in this book, Dennis thinks that larger farms are more viable and the only way that organic agriculture will ever substantially expand: "I think that the reality we face when we look at CSAs or farmers' markets is that most of the time those are very, very small-scale production units, and I really wonder how financially sustainable those small operations are."

"I used to know a guy in the Dallas area who had a CSA, and he lost the majority of his subscribers because he had two crop failures in a row, so all of a sudden he was refunding his subscribers and going without any income," Dennis elaborates. "And if you're one of those small-scale growers, you can't go borrow money from banks, from institutions. They're just not going to take you seriously on a small-scale operation. It's too risky for them. Because if you're small, then most of the time you don't have enough equipment or enough assets to cover whatever operational loans you get. I mean, I spend in excess of $100,000 a year growing my crops, and so there's a lot of capital investment just to have the equipment to get to the point of putting seed in the ground, and then I've got all the growing costs. So that $100K doesn't even take into consideration all the rolling stock and everything else, the whole operation. But without this larger scale, you're not going to see organics have the opportunity or the ability to make major inroads in going from a fraction of the agriculture market to a big share."

Size alone is not sufficient for sustainability, however. Another necessary component is a reliable source of supplemental, nonfarm income. "Family farms everywhere, whether they're organic or conventional, or large or small," Dennis observes, "almost always need another source of income in order to protect and maintain the farming. Land development, putting together subdivisions and owner-financing the sale of the lots, is what I do in what I consider my off-season, from about the middle of June until probably the middle of September. Even though we're basically always growing the next citrus crop, when we're not harvesting and packing, when we've closed the packing shed down, there's a relative calm, so that's a natural time for me to do these other things."

Without this business, which accounts for about a fourth of the Holbrooks' income, and without the harmonious working of all the parts of farming—nourishing the soil, raising citrus, vegetables, and melons, packing and shipping it out—South Tex Organics could not survive. "I've met very few people who didn't have the ability to learn how to become good growers," Dennis says, "but I've met a lot of people who've never learned how to be good marketers and how to stay in business. And for me the main thing has been to have a business that would support the family and provide us with what we desired and what we needed."

Even so, if he could do just one thing, he would do nothing but grow. "If I could do that alone," he muses, "I could be perfectly happy. It goes far beyond just talking about economics. It's a way of life. It's what you're doing every day. It's a stewardship. I've been entrusted with this property, and one of my concerns is will whoever ends up with it, are they going to carry the torch? And that's the reason why I wonder sometimes if I want the business to go outside of the family."

Dennis and Lynda's children, a son and three daughters, are all young adults now. Their son is married and father of a little boy. He and his family live in McAllen, and he works in international banking. The two older daughters and their husbands live in Utah, and so does the youngest daughter, a student at Brigham Young, where all the Holbrook children went. Whether or not any of them will eventually want to work in the business is a question Lynda and Dennis have chosen to keep open. "The business has been good to us," Lynda says, "but we didn't want our son and daughters to be presented with the same option Dennis was without first graduating from college and setting their own directions."

Whatever the distant future holds, no major changes are in the offing for now. The lure of the real estate business is no stronger for Dennis than it has ever been, though he observes that he could make a lot more money in it than farming. "But making money just to make money isn't what fills my cup," he says. Far from scaling back on farming, Dennis is expanding it. In a five-year agreement with the Nature Conservancy as of the spring of 2005, he is leasing 150 acres of citrus at the Lennox Foundation Southmost Preserve, near Brownsville. Prior to purchase by the Nature Conservancy, much of the property was a row crop and citrus operation, and as biologists gradually restore native vegetation throughout the entire preserve they prefer to keep the existing agricultural areas under cultivation rather than let them go fallow. "If you just let the land go fallow," Dennis explains, "there's a tendency for nonnative invasive species to come into the tract, and then they have to battle that in order to do the revegetation. And organic farming is more in harmony with what they're doing than conventional. Better for the soil and water and for helping increase populations of beneficial insects."

The prospect of farming an additional citrus grove that is half again the size of what he is already working pleases Dennis. "It allows us to be on two separate ends of the Valley, with different growing conditions and weather patterns. So it creates new opportunities. Of course it also creates difficulties, transportation, mainly, getting product from down there back up here to our packing shed. But I'm looking forward to the whole thing," he says. "I guess I've just got dirt on my boots and can't get it off."

Contact information

Dennis and Lynda Holbrook
South Tex Organics
P.O. Box 172
Mission, TX 78573

Telephone: 888-895-0108
E-mail: info@stxorganics.com
Web: www.stxorganics.com

Boggy Creek Farm

The Art of Urban Farming

East Austin is home to some of the city's poorest people and richest soils, according to Carol Ann Sayle, and she should know. Since 1992, she and her husband, Larry Butler, have lived at 3414 Lyons Road, raising vegetables year round on five certified organic acres of land surrounding their house. Situated just two and a half miles east of the Texas state capitol—"a brisk, thirty minute walk," Carol Ann notes—Boggy Creek Farm and its neighborhood bear no resemblance to that dressed-up, pink granite world. The streets are lined with small frame houses built in a funky hodgepodge of post–World War II styles, many painted in bright colors. Residents sit on front porches, listening to boom boxes and firing up their barbecue grills. A taquería here, an auto repair shop there, and somehow Boggy Creek both stands out from and yet fits right into this lively community.

An intensively worked farm, it is beautiful and inviting but no more buttoned down and fixed in appearance than neighboring properties. Rows of fruit trees, flowers, and vegetables in the front field extend to within a few feet of Lyons, and depending on the season some rows might be dotted with wire trellising cages or shrouded from insects or the cold and wind by white row cover. And always, mounds of compost are visible from the street, and sometimes also a straw-hatted Larry, operating the small red tractor he uses for turning the compost heaps and tilling the fields. If Larry can't be seen working in the front, then it will be Carol Ann or one of several employees, straw-hatted, too, and sowing, hoeing, or harvesting. A big red tomato-shaped sign to the right of the driveway identifies the farm, much as neon signs mark businesses at other households. Similarly, to the left of the drive, a yellow sunflower-shaped sign indicates that the farm stand is open every Wednesday and Saturday from 9 A.M. till 1 P.M. The drive forms an inverted U between Lyons and the farm stand—a large, tidy shed—and the house, which sit near each other more or less midway between the front field and the back one.

The farmhouse is a four-room, Greek Revival design, built in 1840. "It's an enclosed dogtrot," Carol Ann explains, "with two identical rooms on either side of a central hallway." Made of cypress, it was in near ruins and the land full of junk and weeds when she and Larry purchased the place and set to work, restoring everything themselves. The house now bears a historical marker, and the exterior is painted entirely in white. Its simple elegance is striking, although the house is not an immediately prominent aspect of the farm. Its distance from the street as well as a row of towering, thick-trunked pecan trees marking the edge of its small front yard prevent this effect. So does the inverted U drive, especially on market days, when customers park their

Boggy Creek Farm entrance.

cars at angles all along it, not so unlike Boggy Creek's neighbors do with their own cars in the middle of their own lots every day of the week.

Many customers come not only to buy produce but to linger for a while. They want to explore the fields and see, literally, where their food comes from and show their children. They stroll about, checking out the vegetables and fruit trees and the compost piles and the action in the hen yard. Or they simply sit and take it all in, lounging on benches. Neighbors join this scene, too, some of them customers, some not, since, after all, Boggy Creek on market days isn't strictly a farm but also a park, a good place just to hang out.

Carol Ann and Larry think and plant in terms not of four seasons but two, hot and cool. "In Central Texas," says Larry, a jovial, round-faced man in his late fifties, "spring and fall don't last long enough to count." The hot season runs from May through October and at Boggy Creek features tomatoes, onions, potatoes, garlic, basil, green beans, summer and winter squash, peppers, sweet corn, cucumbers, eggplants, okra, purslane, lamb's quarters, melons, figs, peaches, and pears. The cool season goes from November through April and features root crops such as turnips, carrots, radishes, and beets; kale, collards, chard, sorrel, spinach, arugula, lettuces, escaroles, endives, and other greens; cabbage, cauliflower, Brussels sprouts, and kohlrabi; leeks, green garlic, and green onions; fennel, parsley, and dill; and strawberries.

Like other vegetable growers who sell directly to the public, and especially those who sell year round, Carol Ann and Larry emphasize that a key to commercial success is offering as much variety as possible all the time. "We do over a hundred crops," Carol Ann explains,

Larry Butler and Carol Ann Sayle in the Boggy Creek farmhouse.

"including four to six varieties in certain vegetables, like onions, eggplants, squash, and lettuce."

In addition to the five acres at Boggy Creek, they also cultivate five to six certified organic acres at their farm in Gause, seventy-five miles northeast of Austin in Milam County. They start vegetable seeds at Gause, too, in greenhouses tended by an employee who lives in the area and is a horticulturist. Larry goes there twice a week, to work and truck vegetables back to Boggy Creek. Sales at the Boggy Creek farm stand provide them with about 95 percent of their gross annual income. The remainder comes from selling wholesale to the local Whole Foods store. "Generating most of your revenue through retail is how it should be for all small farmers," Carol Ann says, "if you want to make your entire living by farming."

While Boggy Creek's soil is a medium-heavy clay, the soil at Gause ranges from sandy loam to what Carol Ann describes as "sugar sand" and is particularly good for growing tomatoes, melons, cucumbers, winter squash, potatoes, and onions, which helps them achieve the variety they strive for as well as the volume. Dividing production between two farms also gives them a kind of weather insurance. A tornado struck Boggy Creek in November 2001, for example, and damaged the house and destroyed most of the vegetables, but they still had the Gause crops. Water at both farms comes from wells (though city water is a backup at Boggy Creek when drought makes the well go dry) and is delivered through buried pipes with faucets from which drip tape is run along rows of vegetables. They consider this irrigation method superior to sprinkler or flood irrigation in minimizing runoff and evaporation, thereby conserving water. Drip tape, like row and ground cover, does make litter, Carol Ann notes. "But there are pluses and minuses to everything. And I pick all the litter up, plus I pick up the trash that's blown onto the farm from the road and the trash in leaves that landscapers bring us for our composting." Composting is their main means of feeding the soil at Boggy Creek, while at Gause, with more acreage available for crop rotation, they plant legumes, clovers, and such for green manure.

Along with vegetables they also grow flowers, both for cutting and selling at the farm stand and for the many flower beds near the farmhouse. "This is the most spectacular time for flowers here," Carol Ann tells us on a bright, clear Saturday in early April. And we easily see what she means, for any direction we look we find a profusion of cool-season bloomers. In the back field are a long row of mostly indigo larkspur flanked by kale on one side and collards on the other; a long row of pink and lavender sweet peas climbing a four-foot-tall fence trellis, with brilliant red poppies on either side; and a long row just of red poppies, these so tall and dense that their broad, pale green leaves form a kind of hedgerow. In the front field, between a line of fruit trees and a row of greens, is a wide, riotous strip of orange, white, and yellow snapdragons, red poppies, indigo larkspur, and purple irises. In beds near the house, low clusters of pink, white, and fuchsia spiderwort grow at the base of tall, thick stands of white and yellow irises.

"I want as big a palette of color in flowers as I can get," Carol Ann says. She's an outgoing woman in her early sixties, with long silver-brown hair pulled back from her temples and held in a single clasp, leaving fine bangs to fall onto her forehead. Before becoming a farmer, she was an artist, an oil painter, and ran a gallery and studio in the Oak Hill area of Austin. Some of her paintings hang on walls inside the house: big, yellow-gold flowers rise from black-green foliage in one; in two others, brilliantly white swans are illuminated all over again by their reflections in the water where they swim. Working with color is clearly one of her talents, in farming no less than in her art. "Bold colors break up the green of so many vegetables and make everything more

Inside the Boggy Creek farm stand.

eye-catching," she explains. "And this is as true of the farm stand as the fields and garden beds. Color makes people want to touch things and take them home. So it's good to have as much yellow and orange and red and purple and other strong colors as you can."

Not just color and variety but abundance draws people as well and provides an equally strong guide for Carol Ann and Larry in the presentation of their produce at the farm stand. "Pile it high and watch it fly!" Carol Ann exclaims. "Now that's a sales cliché that really holds up." And so every table top is covered by baskets, buckets, trays, and tubs, each filled beyond the brim with bunches of vegetables and flowers. Young, yellow-green leaves of upright chicories curl over the edges of rectangular trays, for example, while bunches of red and purple radishes bulge from shallow baskets, and

Larry Butler and a worker rest on the back porch of the Boggy Creek farmhouse.

mixed, multicolored bouquets dwarf their buckets and vases. "During market, two harvesters, one being me, are working in the fields the whole time," Carol Ann explains, "bringing more of everything in as it sells, keeping the tables full, the piles high."

The farm employs a total of eight people, three of them full time. Their very first employee, hired in 1995, still works for them, and they have never fired anyone. They attribute their stable workforce both to their urban location—most of their employees live nearby—and to the fact that they have always paid them above the minimum wage for year-round work. "Since we grow year round, we want permanent help," Carol Ann says. "And we understand that everybody's got to live. I mean, who'd want to live under a bridge in Austin, you know, just for the privilege of working at Boggy Creek!"

The path that led Carol Ann and Larry to Boggy Creek and their success in urban farming was a long, meandering one. It started with gardening on their farm at Gause. They purchased the first fifteen acres of what is now a forty-seven-acre farm in 1981, when Larry sold his business in television and video sales and repair. The shop was located in Oak Hill, in the same center as Carol Ann's gallery, which is how they met. Carol Ann, born in Baytown, the second of three children, was raised in San Antonio, where her father worked in the federal civil service. She moved to Austin when she entered the University of Texas and stayed in the area after graduation, teaching Spanish and English before turning to painting. Larry was born in Galveston but grew up in Gause, the only child of cattle ranchers. After high school, he served in the army, doing a tour of duty in Vietnam from 1967 to 1968, then eventually settling in Austin and going into business for himself. "When I got out of the video and TV business, I was burned completely out. And I thought 'I just want to buy a little place in Gause and just be lonely for a while,'" Larry remembers. "My folks were there, and I knew the place and pretty much everybody out there, which is why I picked that part of the world for a farm just to go to and get away."

Larry and Carol Ann put in a little vegetable garden at Gause in 1981, the first year they had the place, but the long stretches of solitude he had hoped for were not to be, and gardening, like everything else at the farm, quickly turned into an off-and-on weekend activity. Married five years by then, and with three children to support—a son from his first marriage, and a son and daughter from Carol Ann's first—Larry soon found himself in the real estate business. "Next thing you know, I was selling houses instead of TVs in Austin," Larry explains. He became a broker, and Carol Ann, with her art market weakening, got into real estate too. But by the mid-eighties real estate took a deep dive, and what was left of the art market went with it. Fortunately, their business skills and Larry's construction expertise carried them through. "I kept my broker license up but mainly what I did was start my own house remodeling and repair business," Larry says. "I worked for this one group of brokers. They'd call me up and say, 'We need windows repaired and some screens fixed and some rotten wood replaced, and a ceiling fan put in and a gas leak fixed,' and so I'd go give it all a look and make a bid, and they'd take it. I was a one-stop shop on whatever they needed to do to get a house ready to sell. They didn't have to call the plumber, then call the electrician, then call the carpenter, you know. They just called me and I went and did it."

Larry recounts these things as we sit in their dining room, a sunlit room with pale yellow walls, where not just the skills he is describing but also fine craftsmanship and artistry are evident: in the longleaf pine floors he restored; in the dining table he made of post oak from Gause; and in the mantel he made from a Boggy Creek pecan, felled by the 2001 tornado. It is Carol Ann, though, not Larry, who identifies these accomplishments as his. "Larry has a lot of knowledge and just plain know-how," she says, "about a lot of things."

Larry gives her a fond, wry grin and says, "Why, thank you, dear," then gets back to the realtors. "They kept me busy from the get-go," he says. "For four to five years I did repairs and remodeling for them and then it just got out of control, got to be way more than I liked, and I got mostly burned out with what I was doing."

That was 1991, and, as he had done during his 1981 burnout, Larry turned to the Gause farm to restore his spirits and somehow lead the way to something better. "I told Carol Ann, I said, 'Well, you know, we ought to grow something out here just to do something different.' I said, 'Why don't we grow some vegetables? And we'll take them back to Austin and maybe we can sell enough to make our little $150 a month land payment. Let's grow some vegetables and see what happens.'"

Carol Ann liked the idea as much as Larry did. "The more we talked," she recalls, "the more we realized that what we really wanted was good clean food for ourselves and some to sell to others." So in the spring of 1991, while they kept their hands in remodeling and real estate, they started spending more time at Gause, and Larry tilled up the biggest garden they had ever had. "We didn't know the first thing about growing cool-weather crops, just hot," he says, "and so I tilled up maybe a quarter of an acre, and we planted onions, potatoes, and carrots." As the weather warmed, he tilled about half an acre more and planted cucumbers, squash, and tomatoes.

"There was no question for either of us that it had to be organic," Larry continues. And even though at that point they were thinking of themselves as gardeners, not farmers or even potential farmers, they got their Gause farm certified organic that same spring. The TDA certification program was still young and cost little then, Larry explains, and except for a few large-scale growers in the Panhandle most certified farmers were small. "In fact, one TDA inspector told me he'd certified a single pear tree for a lady, a pear tree in her backyard, because she wanted to sell certified organic pears," Larry remembers with a chuckle. "Compared with that, I thought I was pretty fair-sized already!"

Driving back and forth from Austin to Gause, Carol Ann read organic-growing books out loud to Larry all the way. *The New Organic Grower* by Maine farmer Eliot Coleman was the first. "I'd say, 'Look at this!' and put some picture in Larry's face," she recalls, smiling, and then, deadpan, goes on. "That book was very inspirational. Very. Coleman said two people could handle five acres. And that was a big lie!"

Larry grins. "Well, maybe in Maine," he says. "Too cold up there for bugs! And in the winter, you can sit and write books and rest up for farming in the summer."

The first big garden at Gause proved a highly productive one, and Larry and Carol Ann enjoyed the harvesting along with the good eating they had looked forward to. But dealing with the surplus bounty was a problem. Like many ambitious novices, they had dreamed of selling and making a little money without figuring out exactly how or where to sell. "We had no idea," Larry said, "but with the crops coming on, we had to do something pretty quick."

They went to their friend Wiggy, owner of the liquor store Wig Liq at West 6th and Blanco, in a little business district about half a mile west of downtown Austin, which at the time also included the Sweetish Hill Bakery. Larry and Carol Ann lived only a few blocks away, in the same neighborhood, and knew what a big business Wiggy's and the bakery did on Saturdays. Wiggy gave them permission to set up a Saturday stand on the side of his store that faced the bakery. "I loaded up my little red pickup and got there before Wiggy's opened," Larry says, "because the bakery opened early, real early. And we had onions, potatoes, and carrots, beautiful carrots, to start, all certified organic, and people really wanted organic, and I made $47 that first day and thought that was pretty good. Before long, we had tomatoes and I was selling $80 to $90 worth, and even though I still had my remodeling business and was still doing a little brokering and knew I could hold on to that and raise vegetables, too, I could tell I didn't want to. I mean, growing the vegetables and then selling them on a Saturday morning, well, boy, it just felt good."

The tomato crop turned out to be so big that he had to find an additional venue. He had planted a couple of

Shed with drying garlic at Boggy Creek Farm.

hundred tomato plants, mostly Romas, though these years later he can't remember why he put in so many, or why Romas. "Maybe I got the seeds cheap or something, I don't know, but we wound up with all these Romas," Larry says. "We took big grocery sacks out to the farm and brought tomatoes back in them. But there were so many, I couldn't sell them all on Saturday mornings." So besides selling outside of Wiggy's, he began taking sacks full to Fresh Plus, a small grocery on West Lynn, and selling wholesale. "But I still had more tomatoes than I knew what to do with."

Whole Foods was nearby, yet when Larry and Carol Ann thought about approaching the store for business, they balked. "Neither Carol Ann nor I either one had the guts to go and try," Larry says. But their teenage daughter, Tracy, did. She took a copy of their organic certification and a sack full of tomatoes and spoke to the produce buyer. "And the guy said, 'Okay, I'll take your tomatoes but not in sacks. You've got to bring them in boxes. Can't bring them in paper sacks,'" Larry explains. "And so we started going there and they'd save us boxes and we'd bring our produce in the boxes." Whole Foods asked what else they could bring, and so did Fresh Plus and their customers at Wiggy's.

Larry tilled up more ground at Gause. "I was forty-seven," Carol Ann says, her voice gleeful. "Larry was about forty-three, and we were having the time of our lives. Didn't have any money! Didn't have anything to lose! Might as well farm!"

Buying Boggy Creek resulted from a kind of accident. In 1992, as they were entering their second year of vegetable growing at Gause, a man Larry had employed in his remodeling business wanted to buy a house in East Austin and asked Larry to help him find one. Since Larry was in the process of concluding all his realty work, he agreed only reluctantly, as a favor to his friend, and he didn't get right on it. But when he did, and opened up his Multiple Listing Service book to find properties for sale in the area, he discovered one that he himself had listed in the mid-eighties. It was 3414 Lyons Road. He had come close to selling it a couple of times, but because one party or another reneged he had not been able to, and the owner eventually took it off the market. Larry had not laid eyes on the place since then or even thought about it. "But that little old photo in that big old book brought it all back," he says. "Even though the picture made it look like just a vacant lot, what with the weeds so high and the house so far off the street, I knew it wasn't. I remembered exactly what was here, and I knew I had to go look, and I wanted to go immediately."

He wasted no time telling Carol Ann and driving them over to Lyons Road. Farming in the city was not something they had imagined before, but Larry's discovery made it seem they were being invited to, and the possibilities they saw excited them. Within hours they were negotiating the price and purchase terms by telephone. The last owner of the place had lost it through foreclosure, and an out-of-town mortgage company held the title. Larry and Carol Ann had no debt, not even credit card debt, but they had no savings or cash reserve, either, and did not want to sell their west Austin house just then. "We had to be sure we could really do what we thought we could do before giving up that house," she says. "So with no money, we were looking at a price of $40,000 for 3414 Lyons, and the only way to do the down payment was max out the cash advances on our credit cards."

They closed the sale in September 1992, rented out their house in west Austin several weeks later, and moved to 3414 Lyons. "It was like camping out for a while," Carol Ann says. "The roof was caved in, the chimneys didn't work, all the fixtures had been stolen, and all the doors, too. It was totally open, just a shell, and thirty-four degrees the first night we spent here." But day by day they grew more certain that this is where they wanted to be, and they soon sold their other house and paid off the note. "It seemed like this place was meant for us to have," Larry says.

They named their new farm after the creek that flowed through it when it was established in 1839 as a fifty-acre homestead. (Long since paved over, the creek ran just south of what became Lyons Road.) They suspended their vegetable growing at Gause and concentrated on cleaning up Boggy Creek, relying on a few real estate deals and a little remodeling for income. By 1994, they had mostly restored the house, received organic certification for Boggy Creek, and were harvesting their first crops on the place. "We had it pretty well filled out in vegetables," Larry says, "and we were gearing up a little at Gause again, and it was all the two of us could do."

They sold to Whole Foods and at a farmers' market on East 7th, just a few blocks away, that the Sustainable Food Center (an Austin organization that promotes local food growing and marketing) opened up about that time. "We'd go over there and sell $30 and maybe $60 on big days," Carol Ann recalls, "and then one day they did a special promotion and brought in bands, and Jim Hightower [then commissioner of TDA] spoke, and we made $120." But the market folded after a year or so, and they decided to try selling from the farm on Saturday and Wednesday mornings. They thought they could attract the small but loyal clientele they had developed at the 7th street market. So Larry made a market table from the for-sale sign that had been up when they bought the place—a table they still use—and they called their customers and several other people, asking them to come and to help spread the word. "We only made $35 the first weekend but we didn't have to leave," Carol Ann says. "We could keep on working, which is worth a lot, because you really do waste a lot of time at a market just sitting around. And little by little, our sales went up."

The next year, 1995, a Wednesday afternoon farmers' market opened in the Whole Foods parking lot, giving them another venue in addition to their farm stand and their wholesale account with Whole Foods. "That was the market that kicked us off really good," Larry says. "We'd run our stand here in the morning and then go over there and set up for Wednesday afternoon and hand out maps to Boggy Creek and talk up our farm stand, and we started getting Whole Foods customers out here."

Another big boost came that summer, when the farm was featured in the lifestyle section of the *Austin American Statesman*. It was Saturday, July 1, and Carol Ann remembers it vividly. "The writer had spent two days out here, and she wrote a long article and there were eight color pictures," she says. "But we didn't know when it was coming out. And when we opened up the paper that morning and saw all those pictures? We didn't even read the article. Oh my God, we said, and ran out and harvested a lot more stuff." They took in $1,200 that day, and even though the figure went down before coming up again the article had an incalculably long-lasting effect. "Four years later, people would still come up and say, 'I read that big article about you a few years ago, and here I am,'" Carol Ann remembers. "It was a really big turning point."

The Westlake Farmers' Market (now Sunset Valley Farmers' Market) briefly provided yet another venue, in 1998, about the time the market in the Whole Foods parking lot was folding. Though Carol Ann and Larry's sales were strong at Westlake, they did not draw many customers from there to the farm, for few were willing to drive from far west Austin. But the farm stand was attracting more people anyway, and Larry and Carol Ann soon stopped going to farmers' markets. By 2000, the farm stand became their sole retail venue and has been ever since, with Whole Foods remaining their only wholesale outlet.

"Whole Foods has always given us great marketing," Larry says. "They give us good displays and do a really good job of identifying our produce as ours, and so a lot of people learn about us that way, and some come here and turn into farm stand customers."

But Larry and Carol Ann know of other small farmers, inside and outside of Texas, who have found it difficult, if not impossible, to sell to Whole Foods, especially after it became a publicly traded corporation and shifted away from small, local producers and increasingly toward industrial-scale growers. And Larry and Carol Ann are also aware of the criticism in the local and national press prompted by this shift. What, then, accounts for their very different experience with Whole Foods? "The farmer has to make it happen," Carol Ann emphasizes. "Everything depends on the relationship between the produce buyers and the farmer, and the farmers are here forever, but the produce people come and go, and we're always starting over with every one of them."

Larkspurs, poppies, and vegetables at Boggy Creek Farm.

"Whole Foods had fourteen team members in the produce department last year," Larry adds, "and we had to get to know every one of them, and we always just hope that some little guy, maybe the stocker, will get to know us a little and our reputation with certain customers and give our produce a good display. And so we just stay on it."

Though some farmers find fault with parts of the national organic standards, Carol Ann and Larry see them mostly as a good development. "You hear a lot about lower standards," Carol Ann says, "but I think they make the standards tighter. You've got to stick a thermometer in your compost pile all the time now, which you didn't used to have to do, and keep enormous records." And Larry points out that the standards give the label "organic" a set of specific meanings that other labels lack. "Like 'natural,'" he says. "I haven't figured that one out. And 'pesticide-free.' And 'sustainable,' that's another one that's too vague. If somebody's selling something certified organic, at least you know what they had to do, or you can find out."

They do, however, think that the record keeping can be unduly burdensome to small, highly diversified growers such as themselves. "For a farm that just does three crops, like a grain farmer, paperwork isn't that much, but we do over one hundred crops and we've got to report where every crop is going to be, every bed, when it's going in, when it's coming out, what replaces it, and so forth all through the year," Carol Ann explains. "Which means about four reports on just one plot for all the different crops. And if you change your plan, you're supposed to file a new report!"

Also, the annual costs of certification keep rising, and this too creates more hardship for small growers than large ones. "We've gone from like $150 to $550 in a year or two," Larry says, "and it'll probably go up another $550 in a year or two, so it'll be over $1,000, and then there are other fees, like a floral license, which is $125 now. They fee you to death!"

Still, they prefer maintaining their certification for now. Not only do they find the standards mostly strong, but without certification they could not get the organic premium from Whole Foods or attract new customers who want to buy only certified organic produce. "Our longtime customers know we'll grow organically whether we're certified or not," Larry says, "but a lot of new ones, people who don't know you, want that label."

They credit Whole Foods for making so many people care about organics, not just in Texas but nationwide. "They've done a tremendous marketing job in the sense that every farmer that's organic owes a debt to them, whether they sell to them or not," Carol Ann explains. "They've made people want organic food, so we've always respected them for that. Really, everybody owes Whole Foods a lot just for creating greater awareness and greater demand."

Larry and Carol Ann see a bright future for small-scale organic farming, certified or otherwise, and have been helping to create that future not just by farming but also through volunteer work. In the earliest years of TOFGA (when it was known as TOGA), Larry served as a regional manager for Central Texas. He and Carol Ann organized get-togethers with other growers and aspiring growers so that people could share information. They have also hosted fundraisers for TOFGA at Boggy Creek and been key panelists in how-to workshops for new farmers at TOFGA's annual conference and other TOFGA events.

In addition, they hold their own educational workshops at Boggy Creek and invite the public free of charge. A popular topic is backyard chicken keeping, featuring the laying flock Carol Ann keeps and frequently writes about in her weekly e-mails to farm stand customers. She has also written and illustrated a children's book about the hen yard and a cookbook called *Eating in Season.* Based on the vegetables she and Larry grow and the meals they cook every day for themselves and their employees, the cookbook is not only a useful, no-fuss collection of recipes but a primer as instructive as any workshop on why it matters to eat local produce in season.

Carol Ann and Larry also pay close attention to legislation affecting food and agriculture, and when action seems necessary they inform and urge their customers to join them in pressuring elected representatives to vote in certain ways, including participating in occasional demonstration marches along South Congress to the Capitol. "So many regulations intended to correct the problems of corporate farming end up punishing small growers," Carol Ann says. For this reason, they, like many farmers, are lobbying against the proposed National Animal Identification System, which has three parts, each entailing fees to comply and fines for noncompliance. One would require every farm where animals are raised, whether commercially or not, to be registered by a state administration agency and identified with GPS coordinates. Another would require every animal, including chickens, whether raised for market or simply as pets, to be implanted with microchips. A third would require farmers, ranchers, and all rural animal owners to report all animal movements to or from the property, including births and deaths, within twenty-four hours of any such movements. The ostensible purpose of the identification system is to enable quick responses to disease outbreaks, such as "mad cow" disease, and to acts of food-system terrorism. "This makes absolutely no sense for small meat and egg producers who sell directly to the public," Carol Ann says. "And the cost and time involved threaten their existence!"

Barring the implementation of such dire regulations as these, however, Carol Ann and Larry see no end to their farming. Certainly, they have no desire to stop. "We love the incredible food and the physical labor and being able to work at home, in so much beauty," Carol Ann says. "And we love our customers, these people who thank us for growing such good food, and inspire us because they love what we do, too! We love it all, still love it after these years, and feel great that we can do it."

The only part of it that has ever seemed like much of a drawback has been giving up travel. "We used to go to Mexico eight times a year," Carol Ann says, a bit wistfully. "But the way we farm means we have hardly any leisure time at all." They also don't have health insurance, though they have reached the point that they are saving a little money and could probably afford it, but so far they prefer doing without. They figure that all the exercise they get and the good food they eat will probably keep them healthy.

"There's nothing we'd rather do than what we're doing," Larry adds. "No way of life we'd enjoy more."

"This is how we define success," Carol Ann says. "It's best to work hard at something, and success is looking forward to doing whatever work it is you wake up every day to do."

Contact information

Larry Butler and Carol Ann Sayle
Boggy Creek Farm
3414 Lyons Rd.
Austin, TX 78702

Telephone: 512-926-4650
E-mail: info@boggycreekfarm.com
Web: www.boggycreekfarm.com

Tecolote Farm

Where the South Meets the West

David Pitre and Katie Kraemer grow 150 varieties of vegetables each year on just five to six acres of their certified organic, fifty-five acre farm, Tecolote. The farm lies fifteen miles east of downtown Austin on the western edge of what once was blackland, tall-grass prairie. Decker Creek flows through the farm, and uncultivated areas include a few live-oak mottes, hay fields, and a pasture for the three young Pitre children's pony, Flash, and two horses belonging to friends. Much of the agriculture that supplanted the prairie is now itself being supplanted by suburban development, and big new houses stand starkly across a road and a yet-to-be-developed field on Tecolote's west. The farm's name is the Nahuatl word for "owl" and was inspired by a pair of great horned owls who lived on the property when the family purchased it, in 1993. A sign with a blue background and "Tecolote Farm" in bold white capitals identifies the place. Posted on the trunk of a tall, old mesquite tree, it marks the turnoff from Decker Lane, a broad thoroughfare, onto Tecolote's narrow roadway, a gently curving one that leads through pasture and then alongside rows of vegetables and finally to a tidy white frame house and packing shed.

Most of the vegetables that David and Katie raise are specialty varieties not available in even the best grocery stores and, by season, generally include not just one type of chicory, say, but as many as four, and not just one kind of bean or squash but three, and so on from one year to the next. David and Katie's delight in food is both deep and wide and suggests something of the venturous eating their customers enjoy. "My whole goal with every type of vegetable, whether it's a beet, a carrot, lettuce, whatever it is, is to find the tastiest varieties that can be grown here and grow them," David says.

He expects even their five-person crew to be genuinely interested in raising the best possible vegetables, and for this reason he hires not migrants but seasonal workers who are pursuing careers in food or agriculture and pays them as much above minimum wage as his profit margin will allow. These are mostly men in their twenties who already live in the Austin area or are drawn to it as word of general and specific opportunities circulates within their social and professional communities. A current worker, for example, who previously did farm work in Maine, Illinois, Mississippi, and Oregon, moved to Austin to be with a girlfriend and learned through the grapevine about seasonal work at Tecolote. Most crew members, though, have little if any farm experience, and training a new crew every year takes time and effort. "But it's great because most of the kids are excited about this kind of work," David says, "and we have a lot of fun. Each crew has its own personality. Last year it was all about politics. Just nonstop political debate all year long. Yelling at each other out in the middle of the field! It got pretty funny sometimes! This year we have a music crew. They argue about musicians."

During the fall and winter months when, by choice, David has nothing to harvest or market, he studies seed catalogs for new varieties to plant and test for cultivation requirements, productivity, and taste. "The red beets we're growing now are the best red beet I've ever found." Other recent finds include an Italian head chicory, several French butter lettuces, Japanese radishes, and Korean cucumbers.

David and Katie sell their vegetables mainly through a community-supported agriculture program, which they established in 1995 by sending flyers to alternative health-care professionals. In the Tecolote CSA, subscribers have the option of paying fully in advance or in several installments and receive half-bushel baskets for eighteen to twenty weeks, from March into August. Many CSAs designate certain sites and times for pickup, but the Tecolote fee includes weekly delivery to the customers' doors on a Monday, Wednesday, or Friday morning.

The Tecolote CSA is probably the first CSA in Texas, and now the oldest. Its membership has increased from the original thirteen subscribing households to 150, with 300 more on a waiting list. "We price our baskets to be accessible to grad students as well as CEOs," Katie says, "and we really do get the range. It's great!" In addition to the CSA, they also sell at the Austin Farmers' Market and the Sunset Valley Farmers' Market from March into October, concluding their marketing for the year when such heat-loving crops as okra and eggplant finally give out.

Though David and Katie like the connections with people that the farmers' markets give them, their hearts are much more in their CSA and the even stronger ties they enjoy with their subscribers. "Our subscription lets people experience what vegetables are supposed to taste like, and once did taste like," David says. "I think of our CSA as a course in eating seasonally and learning new vegetables that mostly aren't new at all, but old, rediscovered heirlooms. And I love knowing who's going to eat our produce, and being able to picture their faces even when I'm planting or pulling vegetables. That's the most important part of it for me. And it keeps me honest, because I'm not going to send out anything I wouldn't eat myself." Katie shares these sentiments: "I love how much our produce means to people who can't grow their own food. I didn't really appreciate this when we started, but I've since learned and come to love how much our customers value what we do. Connecting with them, I know why we do what we do and feel it's all worth it."

They celebrate this connection each year by hosting a spring potluck for their subscribers. Our visit, on a beautiful Sunday in late April, coincided with this event. Arriving early that morning—a clear one, and cool enough for a jacket or at least long sleeves—we walked and talked with David and the field hands as they harvested for Monday's fifty CSA deliveries and stacked the produce in crates in the back of a small truck. More work followed in the washing and packing shed, where our conversation continued into the early afternoon. After that, we joined Katie in the house and helped prepare food and fresh lemonade for the party. The mid-afternoon arrival of guests took us back outside, to the live oak–shaded yard and a view of the vegetable field where we began the day, the source of most of the dishes lining the long, cloth-covered tables: beet salads, fennel salads, and salads of mixed lettuces, chicories, arugula, and radicchio; pastas tossed with sauteed garlic and chard or other greens; dips of herbs, onions, and garlic; and farmer Katie's casseroles of red beets seasoned with fennel—huge, steaming hot, and sweet to smell.

Tecolote Farm vegetable field.

The afternoon was pleasantly warm, and many people wore shorts and sandals. Some brought chairs, and others brought blankets and sat on them to eat. The group, which probably totaled three hundred before it was all over, included singles, couples, and families, with ages ranging from a few weeks to seventy-some years. A young mother with a baby wore a bright pink T-shirt that had "Keep Austin breastfeeding" printed on the front. And there was a child or two in tie-dye, and a graduate student with a nose ring and several studs on each of her ears. Most people, though, came nondescriptly dressed, the conventional-minded blending in with the bohemian, the more affluent with the less. Only rarely and incidentally did conversations turn toward occupations and reveal, for example, a young Mexican American man to be a physician, a young African American woman to be a nutritionist, and a middle-aged white woman to be a computer executive. Sharing food at the farm where it

David Pitre harvests lettuce.

grew kept most of us marveling at the tastes and textures, and trading recipes and cooking tips. "What we get in our weekly baskets is always changing!" one of the subscribers told me. Another chimed in, "I didn't know there were so many kinds of beets in the world, or greens and garlics and leeks! It's just amazing, and makes cooking really interesting and fun, not to mention eating!"

Despite their customers' love of good food, however, David and Katie have observed that their attitudes on human and ecological health vary. "We've got a lot of people who are very health conscious and very concerned with eating things raised naturally, sustainably, without pesticides and other things that have an adverse effect on the environment," David explains. "But we have just as many people who really couldn't care less about the environment or poisons. They just want the best-tasting vegetables they can get." David himself does not separate the two. He is convinced they go hand in hand. "I feel very strongly that you cannot have good-quality, good-tasting, healthy vegetables that are grown in a way that is not sustainable, that's not healthy for environment. You just can't do it. They're not two separate issues."

Many of their health-conscious customers value organic certification as assurance that the vegetables they purchase are nontoxic, and this is especially true of their farmers' market customers, who are less familiar with Tecolote than CSA customers are. But each year at certification renewal time, David debates whether or not to recertify. Using organic farming methods that he considers more beneficial for the soil, more genuinely sustainable, than the national standards require is much more important to him than organic certification. He thinks that many of the standards established in the national program promote corporate, industrial-scale organic farming which, much like conventional industrial agriculture, is motivated by economic profit to the detriment of ecological and human health. "The vast majority of organic products on grocery shelves are grown by huge corporate farms, and a lot of these are growing a field of the same thing conventionally right next to an organic field. They're growing organic because there's a buck to be made. And they're growing the same varieties essentially the same way they do conventionally, minus only a few of the pesticides. And the result isn't healthier soil or better-quality produce, even if it is certified organic."

David also objects to the use of plastic mulches and drip tape systems, which are permitted by the national program. Plastic mulches, in his view, do nothing to create healthy soil but harm it instead, as tatters and residues remain and gradually break down. Plus, even if mulches are retrieved after use in fields, they amount to so much expensive waste. Drip tape systems, too, usually entail throwing away huge amounts of cheap, flimsy plastic tape every year. Neither use of plastic seems to David a legitimately "organic" or "sustainable" practice. At Tecolote, an aluminum pipe delivers water to PVC pipes running down each row of vegetables, facilitating both the sprinkler irrigation David uses in spring and the drip irrigation he uses in summer. And David is constantly feeding the soil. He does this by using a compost of turkey litter and rice hulls and by alternating food crops with such cover crops as legumes and ryegrass, which are tilled under as a green manure before the next planting of vegetables. "Fertilization means feeding the soil, not the plant," he emphasizes. "A good vegetable is only going to come from good soil."

Loving good food lured David into farming. He grew up in a family obsessed with food. "I'm not joking when I say 'obsessed,'" he says. "My family! At breakfast we're talking about lunch, lunch we're talking about dinner, dinner we're talking about tomorrow's dinner." David's father is the first male in four generations of his Cajun family to live past the age of fifty-three or fifty-four. "They all died of heart disease," David explains, and speculates that genes played only a minor role, if any. "They ate nothing but sausage and duck. Just fat, fat, fat." It's a propensity that goes with a certain territory. "People from Jennings, Louisiana," David says, "can't seem to stay within thirty-five pounds of their ideal weight. Whole town of fat people!" His tone is more fond and wistful than disapproving. David's father hasn't lived in Jennings for forty years, but he is still overweight, and David, who never lived there at all, grew up eating as if he did. "I was pretty hefty my whole life," he says, "till I moved away from home."

A hefty David is hard to imagine, however. He is lean and fit and, though blue-eyed and fair-skinned, also well tanned—a picture of health and active enjoyment of the outdoors. Farming from college on can do that for a person. "I like hoeing. I like pulling weeds," he says. "I love being outside and getting my hands dirty!"

Born in Mexico City in 1964, David was raised along with a younger sister in Richardson, Texas, which he recalls as "country" when he was growing up, though it

David Pitre and workers wash and ready vegetables for CSA boxes.

Henry Pitre and a worker take a break; greenhouses in background.

has long since become part of Dallas. He attended the University of California at Santa Cruz, where his interest in food and the natural environment led him to major in agricultural ecology. It was there that he eventually met Katie, who was born in Riverside, California, in 1968. After graduating from UCSC, David worked for several years on organic vegetable farms in northern coastal California. Then, early in their marriage, taking advantage of their youthfulness and lack of encumbrances, he and Katie moved to Palmer, Alaska. David worked on a big carrot and potato farm while Katie, a bilingual educator, taught in the elementary school. They lived in Palmer for a couple of years and really enjoyed it. "But we got pregnant and decided we didn't want to have a baby in the Alaska winter," David explains. "Katie wanted to move back to California, and I wanted to move to Louisiana or somewhere in the Deep South. We compromised on Austin, where the South meets the West, you know. And we came here with the intention of buying a farm and raising produce and selling it retail."

Their parents didn't think David and Katie's plan was the most prudent choice and cautioned against it. David's father, an artist who became a florist when his painting career ended, wondered if David knew what

Katie Kraemer prepares for the Tecolote CSA potluck lunch.

he was doing and, along with David's mother, feared that David and Katie would not be able to make ends meet. For the first couple of years, they did work at a loss, and they didn't consistently break even until about the fourth or fifth year. David describes their income as middle class, noting that their idea of middle-class living is probably lower than that of many people. They don't drive new cars, wear expensive clothes, have cable television, or buy sporting equipment. They also don't have the best health insurance. "Farmers don't make enough to buy it," Katie points out. She and David insure themselves with a high-deductible policy that is also an IRA and the children through the state-subsidized Children's Health Insurance Program.

"We can live on not too much," David says, "and now my parents are very excited by what Katie and I've done. But it took a while for them to come around. I sort of had to prove myself." Even Katie's father, a farmer himself, an orange grower and a cattle rancher near Riverside, had his doubts. He had never encouraged any of his children to go into agriculture. "But you know," Katie says, laughing, "when he comes, he loves taking part in whatever we're doing, though he still thinks it's a crazy life. You should've done something else, is what he thinks."

Katie's upbringing didn't prepare her to be a farmer any more than David's prepared him. "You can leave trees and cattle for a while," Katie says, "but not vegetables." And not only that, but the family—a Catholic family descended from Spanish land grantees, in which Katie is the sixth of seven children—lived neither on the ranch or farm but in a house in Riverside. Katie's daily life was urban, not rural. "Our lives were more parallel to other kids we knew than my kids' lives are," she explains. "My dad left the house for work each day, and my mom took care of us and was always really involved in our school and church activities. And although we were all proud to be a farming family, partly because it was already becoming a rarity, I wasn't a country kid."

By contrast, Katie and David's children, Zach, twelve, Claire, nine, and Henry, six, have lived on Tecolote since birth. From 1993 until 1998, the family lived in a two-bedroom trailer, intending eventually to build a house. But by the time they thought they could build, the price of lumber had doubled. They were determined, though, to get out of the trailer. "It was terrible," Katie says. "The floor had holes in it." She searched newspaper ads for old houses to be moved and found a six-room Victorian for $8,000. They purchased it and, at a cost of $15,000, moved it to the farm. They are now restoring it, but gradually. Previous owners had put lemon-and-lime colored paneling on every wall in the house. "Ugly," Katie says, "hideous." So, room by room, they are redoing the walls. Many ceiling panels, too, have had to go because of paint flaking off. For the kitchen-and-dining room, the largest room in the house and the main gathering place, David installed salvaged pressed tin. Not until 2004 did they get central air conditioning and heat. Before that, they relied on fans and a wood stove.

"Our kids have had a crazy childhood because of our redoing the house and farming," Katie says, only half joking. "The older the kids get, the more aware they are that our lives are kind of different, especially the oldest, Zach." This awareness and the discomfort it sometimes causes might not be so sharp, Katie thinks, if the children attended public schools near Tecolote, but they don't. Katie and David, despite favoring public education in principle, find that the local public schools are not as academically challenging as they would like. They send their children instead to a private school in Austin, aided by partial scholarships.

Though students at the school come from diverse economic backgrounds, they are all urban or suburban. When schoolmates come to the farm, whether on personal visits or class fieldtrips, the Pitre children sometimes feel self-conscious about living in a house that is nothing like some of their schoolmates' big and fancy homes full of electronic games and gadgets. Many of the kids don't like to come to their house, Zach occasionally observes to Katie, because they get bored. "But Zach," Katie will say, "isn't that sad? With so much to do out here?" And, in moments like these, she often asks him if any of those other children have acres and acres of space for play, a creek and a treehouse, horses to ride, and chickens for eggs and goats for milk. "And he does recognize," Katie says, "if he stays too long at somebody's house and just sits inside with games how bored he gets and how ready he is to come home and do things he can only do here." Still, she admits, it is hard for her that her children feel this split.

Like David, though, Katie loves farming, loves working outside. "I'd always rather work outside than in my house," she says. "I'd always rather be out with the crew than on the computer." Katie is tall, with dark brown hair and eyes and an olive complexion that is deepened, even in April, by days of sun. During the children's toddler years, she spent less time in the field than before they were born, but since all are now school age, she is in the field quite often, though not on a regular basis. "I'm in the field when I need to be, when we're short or when we have a lot to harvest." Besides frequent field work, Katie is the one who milks the two Saanen goats twice a day and tends the motley laying flock of Barred Rocks, Araucanas, Andalusian Blues, and Rhode Island Reds. And in December she works with David in their greenhouses, starting vegetable seeds that they will later transplant into the field for spring and summer crops.

Much of Katie's work, however, takes place inside. One of her largest tasks is managing the CSA. This includes not only keeping accounts but determining the routes for the 150 weekly deliveries, fifty each Monday, Wednesday, and Friday during the eighteen- to twenty-week subscription period. She and David hire a driver, but to train that person and test the efficiency of the routing Katie does the first week's driving herself. Katie also composes and prints weekly newsletters for subscribers, which they receive along with their vegetables. The newsletters include information about the vegetables and some of her and David's favorite recipes—garlic soup, sorrel salad dressing, parsley and fennel salad, peppery Texas pickles, and stir-fried long beans, just to name a few. Katie sends recipes to customers by e-mail, too.

Though most of Katie and David's income is from the farming, Katie is employed part time from September through November by a major commercial publisher. Working at home, she edits textbooks for factual accuracy. She enjoys the work. It gives her some of the intellectual stimulation she previously got from teaching, something she would like more of, but later. For now, farming and raising a family present plenty of challenges.

Just to have much of a family life is, in fact, a big challenge. "It's the hardest thing," Katie says. "There's just no time." So far, they have managed to take a family vacation each August, when the CSA subscription ends for the year. They usually vacation with relatives, joining members of Katie's family in Newport Beach or Lake Powell, or David's in Mobile, St. Louis, or Santa Fe. But the increasingly early start of school has made a vacation of any length difficult. Still, August and the fall months do allow more time for being together as a family—more time for preparing and sharing meals, for giving and going to parties, for movies and eating at special restaurants, and for simply relaxing a bit more often at home.

The "off season" at Tecolote is, however, hardly free of farm work. David and Katie harvest heat-loving vegetables and sell them at weekly farmers' markets into October, and David plants garlic and onions and sows cover crops for green manure. He also repairs machines and buildings, having taught himself these skills in the course of farming. A farmer has got to be handy, he thinks, or the result is hardship on just about everyone involved. In addition, David studies seed catalogues and determines which varieties to grow and when to plant them, an intricate and critically important task when you raise a 150 varieties, many with multiple plantings. "And in an unpredictable climate!" David says wryly. "When I came here, I thought I knew all about farming because I'd worked on so many California farms, but I really had to start from square one." Though the vagaries of weather are always a challenge, the knowledge David has gained from years of diversified cultivation is a boon. By January he has created his planting calendar for the new year, he and Katie are starting seeds in the greenhouses, and Katie is working on CSA renewals. Whatever extra time they had for special family activities or as a couple has gone, the business of farming once again taking most of each day. "Basically, in January, David says if anybody invites us

Tecolote CSA members enjoying potluck at the farm.

over, tell them I'm busy," Katie explains. "From January to August, we don't go to parties or do much as a family. We're just too tired. It's very taxing on your body—twelve-hour days almost every day."

David especially feels the physical toll. He doesn't take a day off from mid-February to the first week in August. "And a lot of my days are eleven- to thirteen-hour days," he says. "By August first, as often as not I've worked myself into such a state of exhaustion that I can't get out of bed for several days." At the age of forty, he sometimes feels fifty or sixty and wonders if holding up to the work for twenty more years or so is even possible.

He also wonders if he really wants to. "I think about it almost daily," he admits. "Something happens just about every day and I'll think, 'This is not working! This really sucks!'"

He has many interests and, in these moments, dreams of doing all kinds of things. Becoming an academic, maybe a food anthropologist. Or maybe doing a different type of farming, something less physically exacting than raising vegetables. For a while, he thought about raising frogs and producing frogs' legs, and then he thought about raising ducks and producing meat from them. Recently he has been thinking about raising pigs on pasture for pork. He and Katie love pork. "We're porkaterians!" he laughs.

He and Katie have also considered taking a whole year off somehow, getting some sort of fellowship and going with the children for a year of study in France. "Small-scale farming is much more established in Europe than here. So we're kind of wanting to go and learn what machinery's being used and what techniques, just to make our own farming more efficient," Katie says. A while back, they purchased an Italian spader which, unlike rototillers available in the United States, does not invert layers of soil. They also took a class in French but found it daunting, though both are fluent in Spanish. A year in France may be just a pipe dream, Katie says with a shrug and a smile.

"The life of farming," Wendell Berry has observed, "can be for the same person at the same time close to unbearable and yet irresistible." For now, Katie and David have no real plans to change what they are doing. "I really, really love this," David says. And Katie, given her druthers anytime and anywhere, would be outside, growing perennial flowers and, if David weren't doing it, vegetables too. "That's what I would do," she says, "all the time." Somehow dreams of escape keep them rooted in the dark, fertile dirt of Tecolote.

Contact information

David Pitre and Katie Kraemer
Tecolote Farm
16301 Decker Lake Rd.
Manor, TX 78653

Telephone: 512-276-7008
E-mail: tecolotefarm@juno.com

Animal Farm

How Gita Vanwoerden Accidentally Became a Vegetable Farmer

G**ita Vanwoerden has been** growing vegetables commercially since 1992. She sells her produce to some of Houston's best chefs and at farmers' markets in both Houston and Austin. When she became a rural landowner, however, she did not have vegetables in mind but weekend recreation for her family.

In 1991, Gita, a native of what became Israel, and her husband, Cas, originally from Holland, set out to find a farm within an hour's drive of their home near Rice University, in central Houston. Cas, an electrical engineer, owns and operates a business in which he designs and builds control instruments for oil and gas installations. His work had brought them to Houston from Amsterdam ten years before, when their first child, a daughter, Dana, was five. Gita, with a degree in art, occupied herself with oil painting in addition to homemaking and child rearing. As their family grew, with the birth of a son, Adon, in 1986, and another daughter, Salome, in 1988, Gita and Cas found themselves wishing for a rural retreat. They wanted somewhere just to relax and ride horses. Gita had already purchased a horse, an Arabian, the breed she considers the fastest and most intelligent, and stabled it at nearby Hermann Park so that she could take the children often and teach them to ride. But she did not like the quality of care the horse got or the limited terrain for riding.

Gita and Cas thought a twenty-five-acre farm in the Brenham area, seventy miles northwest of Houston, would be the right size and the right distance from the city, but to their dismay they could not find anything there they could afford. Returning to Houston one Sunday afternoon from yet another discouraging search, they happened to drive through Bellville and stop at a realty office on the edge of town. A seventy-acre place near Cat Spring, just sixty miles almost due west of Houston, was listed at a cost far less than the smaller farms they had been looking at, and it even included a spring-fed creek. The Vanwoerdens went immediately to take a look and found, at the end of a winding three-and-a-half-mile dirt road, an open gate but only a foot path, no road, into the property. "It was getting toward evening, and all we could see was this sandy, overgrown path into a thicket of cedars and oaks," Gita says. "We thought the woods were beautiful, but it cost so much less than the other places we'd looked at, we were afraid something might be wrong with it."

To see if they could dispel their misgivings, they got permission from the owner, a teacher in Bellville who had inherited the property, to return on subsequent weekends. She allowed them to bring machetes and

Weeding a vegetable plot at Animal Farm.

loppers and make the path passable enough to explore the place as thoroughly as they wished. No one lived there, or ever had, so there was no house or even a well. The owner had always leased the place to hunters and decided to sell it when she could no longer lease it at the price she was accustomed to. But after listing it for months, she found she couldn't easily sell the place either. Access seemed to be the main problem. "The sandy, winding road put most people off," Gita explains. "And by the time we looked at the property, the owner was feeling desperate and had lowered the price just to get rid of it."

The more the Vanwoerdens explored the farm, the more they liked it—not just the woods but the single small meadow in the interior and the swimming holes in Sandy Creek—and the more assured they became that nothing was wrong with it. So in June 1991 they purchased the property, had a well drilled, and hired a neighbor to make a road into the place. They had a corral and horse barn built and bought three more Arabians to bring out with the one they already had. They also purchased a couple of Boer goats for milk and breeding pairs of ostriches and emus and settled them onto the place, too. "We thought that selling year-old birds would help us pay for the farm," Gita explains. "But we lost interest in that and just kept the birds for pets until they died."

Populating their farm so quickly with so many animals made Gita and Cas think of the George Orwell novel *Animal Farm*, despite its being a grim fable of a Soviet-like totalitarian state. "Initially, I wanted to name our farm Sandy Creek Farm because of the creek and because the soil is so sandy," Gita says. "But then I thought, no, I just really like the name Animal Farm, because that's what our farm will be, an animal farm, but a different kind of animal farm."

Gita is short and small and has dark hair and big, black-brown eyes. She speaks English with a British accent, and it is not her first language but her third. Born in Jerusalem in 1947 and raised there, she grew up speaking both Hebrew and Hungarian. Her mother, a cook for the British mayor of Jerusalem, was a Hungarian Jewish émigré. Her father, an accountant employed by the British army before and after the creation of Israel, was a British Jew who was born in Java, Indonesia, and educated in Singapore and England. He spoke Dutch and Hebrew in addition to English, and her mother, in addition to Hungarian spoke Hebrew and English. But her parents did not speak English with Gita and her older sister, just Hebrew and, in their mother's case, also Hungarian. Hungarian was reinforced by their maternal grandmother, who lived with the family and took care of the girls while their parents worked. "English my parents reserved for discussing things they didn't want my sister and me to understand!" Gita says, with fond amusement.

She did learn a little English in high school, since it was a standard part of the curriculum, but she did not begin to develop any fluency until she went to Johannesburg, South Africa, to attend the Johannesburg Fine Arts College. College in Johannesburg was her mother's idea. Her father had died of heart disease a year before Gita graduated from high school, and her sister, eight years older than Gita, had been living in Johannesburg since soon after marrying. "I had an American boyfriend whom my mother did not like. She thought he was a crazy cowboy!" Gita says, laughing. "And so when I finished high school, she called up my sister and told her she was sending me off! And I was very glad, really. I had spent the summer before with my sister and I really liked it, and had checked out the art school."

She graduated from college in 1973 and, before starting a job teaching art to adults at a private school in Johannesburg, traveled with a girlfriend to Europe. In Amsterdam, she visited a high school friend who had moved there, and the friend introduced her to Cas. The same age as Gita, and one of four children of a Dutch Reformed pastor, Cas grew up in a village near the Holland-German border. Having recently received his engineering degree from a Dutch technical institute, he was living in Italy, working for an Italian oil-and-gas control instrument company, and happened to be in Amsterdam on business when he and Gita met. They were smitten with each other. In order for them to be together, Cas quickly arranged to open and supervise a Johannesburg office for his Italian company, and two months later he and Gita were married. For the first three years of marriage, Gita, along with teaching, also painted—oils of large, abstracted human figures—and exhibited and sold her work. In 1976, Dana was born, and five months later, because of the social and political turmoil related to the dismantling of apartheid, the Vanwoerdens left South Africa for the Amsterdam area and lived there until moving to Houston five years later.

Gita grows vegetables year round on ten noncontiguous plots that total eight acres. She calls them gardens, and they range in size from 50 by 50 feet, a small fraction of an acre, to 200 by 300 feet, an acre and a half (an acre being 40,000 square feet). The two earliest gardens are the shape of a circle and a triangle, but as she developed her markets and expanded production Gita quickly found rectangles more practical, and so the later eight gardens are rectangles of varying dimensions. Sculpted one by one from the woods during the course

Plastic-covered hoop house at Animal Farm with young lettuce.

of fifteen years and bounded by thicket, the gardens cannot be viewed all at once. Walking from one to another, although you know you are on a farm, you sometimes feel you are hiking through a forest and being surprised by glades that turn out to be not wild, uncultivated clearings but tidy, flourishing food gardens. Cool, quiet gardens, or so they were on the late March day of our visit, a soft day, thinly overcast, still and peaceful.

Gita planted her first vegetables during the family's first spring at Animal Farm, the spring of 1992. The Vanwoerdens were spending every weekend at the farm by then, from Friday through Sunday night, sleeping in a travel trailer they bought for temporary housing. They were enjoying swimming and riding, taking care of the goats and emus and ostriches, and walking through the woods with their pet monkey and two pet dogs. Plus, a house was in the works—Cas and a friend were beginning to build an off-the-grid farmhouse. The only thing missing was a vegetable garden, Gita thought. "Salome was about four, Adon, six, and Dana, sixteen, and I thought it would be a good experience for all of us, working together in the garden and growing some of our own food," Gita says, noting that the family had become vegetarians a couple of years earlier, and remain so. "I just wanted a small kitchen garden and didn't expect anything beyond that to come of it, and the only thing I knew was that I didn't want to use chemicals."

She had had a garden before, in South Africa, but she had not kept it herself and had learned nothing about growing vegetables. "We lived outside of Johannesburg and had excellent soil and a nice, big garden. And I remember these huge cauliflowers in particular. They were beautiful and delicious. Everything was. But the gardener did all the work!" she explains, laughing. "A lot more than I did! All I did was purchase some gardening books to read, but that was really silly. The books were from England, an entirely different climate."

To grow vegetables organically at Animal Farm, Gita realized she should seek information and advice relevant to the local climate and the farm's sandy soil. Looking in the Houston telephone book, she discovered Garden-Ville (a San Antonio–based business specializing in organic gardening products founded by Malcolm Beck in 1957), then visited a Garden-Ville store in a strip mall in Stafford, a southwestern suburb of Houston. Mark Bowen, a professional horticulturalist and author, ran

Workers at Animal Farm sort harvested vegetables.

the business at the time and was immensely helpful to Gita. "I told him I had absolutely no idea what to do, and he really coached me," Gita says. "He told me what fertilizers and mulches and other soil amendments to buy for my sandy plot, and what vegetables to start with, and how to make compost from our animals' manure and from kitchen and plant waste. He was wonderful!"

Gita started with several varieties of tomatoes, cucumbers, and squash and returned to Bowen's store frequently as she cultivated her garden and fed the soil. She never wore out her welcome with Bowen, she says, not only because he was generous in sharing his expertise but also because Gita couldn't go anywhere in those days without their pet monkey on her arm—a small, white-faced capuchin named Buddha—and Bowen loved getting to play with him whenever Gita came to the store. "Buddha thought I was his mother because we got him when he was only five days old, and he was always terrified to leave me. So we put him in diapers and, wherever I went, he was attached to my arm or my neck," Gita explains. "And Mark would keep me in the store as long as possible and chat forever because he didn't want Buddha to go. So he would ask me everything about my garden—how this and that were doing—and explain to

Gita Vanwoerden and a young tomato plot at Animal Farm.

me what to do about any problems and how to prepare for gardening in the fall."

Bowen was such a thorough teacher and Gita such an avid student—"I couldn't believe how much I loved growing vegetables!" she says—that before long, the only problem she had to report was too much produce, especially zucchini. "I had no idea a few plants in a small space could produce so much." Gita says. "The zucchinis were especially productive, and after about two months straight of zucchini, Cas and the children said, 'All right, no more! We don't want to even see it! Let it rot! Turn it into compost!'" But Bowen had a different idea. He put Gita in touch with a landscape designer who kept a big vegetable garden in Houston and sometimes sold his surplus to Jimmy Mitchell, the noted former chef of Rainbow Lodge. The landscaper began picking up Gita's produce every Monday morning when the Vanwoerdens returned to the city and selling it for her along with his.

But soon, when his surplus ran out, Gita delivered hers to the restaurant herself and met Mitchell for the first time. "He said, 'Why are you bringing me so little? I could use a lot more. Could you grow such and such for me? I could use as much as you can grow,'" Gita recounts. "And some of the things he wanted I had no idea about, but I said, 'Sure! I'll try!'" Mitchell became one of Gita's best customers and remained so for ten years, when Rainbow Lodge changed ownership and he left Houston. "He would visit the farm with his friends every two or three months and always encourage me to grow more of certain things I was already growing and to grow new things," Gita says, "and he told other chefs about me."

One of these was Monica Pope, who called Gita in 1994. Though several of Houston's best chefs cook with local produce, Pope, owner of the restaurant t'afia, is perhaps the best known for doing this, and she is one of Gita's main clients. Gita also sells produce regularly to Quattro, the restaurant at Houston's Four Seasons Hotel, and to Mark's, Brennan's, and 17.

Her crops each year include many varieties of several vegetables: at least six varieties of beans; five varieties of beets and carrots each; more than twenty varieties of tomatoes; thirty to forty varieties of lettuces; at least two dozen different greens; several varieties of broccoli, cauliflower, and other brassicas; half a dozen different eggplants, peppers, squashes, cucumbers, and melons; four varieties of onions and leeks; two or three varieties of corn; and a couple of varieties of Jerusalem artichokes.

Since 2003, with the development of new farmers' markets in Houston and Austin, Gita's market has greatly expanded, extending from chefs to a much wider public. Assisted by Cas or a farm employee and, when their college schedules permit, also Salome, who attends Rice, and Adon, a student at Vassar, Gita sells vegetables Wednesday afternoon and Saturday morning at both the Bayou City Farmers' Market (in Houston) and at the Austin Farmers' Market.

Gita enjoys the expansion, and not just because it gives her a larger, steadier income but also because it makes her work more challenging and interesting. "The upscale restaurant business is not a steady business. During the summer it's pretty much dead because a lot of restaurant goers travel, so when I sold just to chefs I always cut down my summer production and kind of slacked off in my farming methods for several months of the year," Gita explains. "But having to show up with plenty of produce for Wednesday and Saturday markets every week helps me work almost constantly to improve my methods, to accommodate the demand and keep my volume up."

She has learned, for example, to make extensive use of large hoop houses—she has five of them, each measuring 96 by 20 feet—to lower the temperature or raise it, as necessary to extend the growing season for some of her crops. (Except in the seed-starting stage, all of her crops grow in the ground, however, not pots, whether sheltered by a hoop house or not.) The late March day of our visit, she showed us into a hoop house where pepper and tomato plants survived a winter that brought several nights of twenty-degree lows. She brought them through not only by extending the plastic cover to the ground but by wheeling in a "chicken tractor," or portable house, of about ten laying hens. "Though they're small creatures, they generate a good bit of body heat," Gita explains, "and containing it keeps them warm as well as the plants."

Another hoop house protects hundreds of young cauliflowers, specialty cauliflowers, some with gold heads, some with yellow-green, and all so tiny yet that you have to peer into the heart of their leafy cauls to see them. As summer approaches, Gita replaces the heat-generating plastic cover of the hoop house with a black meshlike shade cloth, consequently extending her cauliflower season for a longer period than if she grew this vegetable only in an uncovered plot. The growing season for lettuces, however, is the one she has learned to extend the most. Using both shade cloth and a misting system, she is generally able to take lettuce to market every month except August. "But then in August, we're planting lettuce already again," Gita notes, "so really I grow lettuce all twelve months and market it eleven." In the brutal heat of Texas, this is quite a feat.

To feed her soil and keep it healthy, Gita follows, in each garden, a sequence of crop rotations that she adapted from Maine farmer and author Eliot Coleman. Her version has eight parts: beans are followed by tomatoes, eggplants, or another member of the nightshade family; nightshades are followed by peas; then brassicas or greens are planted; next, corn or a cover crop; then potatoes or a cover crop; then melons or cucumbers; next, a root vegetable such as carrots, onions, or beets; and then back to a bean again. Her main cool-season cover crop is a mix of oats, rye, clover, and vetch. For warm-season cover crops, she sows buckwheat and black-eyed peas.

In addition, she makes her own compost and applies it, as needed, throughout the year. "We use vegetable waste from here on the farm, plus weeds, and vegetable waste from some restaurants and from farmers' markets," Gita says, "and we monitor the temperature in our heaps and when it reaches 140 degrees, we turn the pile with a front-end loader. Generally we're able to produce good compost in six to eight weeks. The key is monitoring the temperature and turning the stuff often."

Organic certification through TDA is in the offing, pending only the agency's completion of the paperwork. "When I sold just to chefs, I didn't need to bother with certification," Gita says. "They all know me and know for sure that I use organic methods, but that's of course not true of customers at a farmers' market, and I want certification so that people who don't know me can be assured that I'm organic."

Maintaining a stable workforce on a farm that is somewhat off the beaten path is a problem, but supplementing employees' pay with on-farm housing, which Cas built, plus all the produce they need for food, helps mitigate it. Currently, Gita employs four full-time workers, and she works more than full time herself. She does all the planning and seed starting and works in the gardens, cultivating or harvesting, most of every Monday, Tuesday, Thursday, and Friday. Tuesdays and Fridays, in the late afternoon or early evening, she goes to Houston for the night, to deliver produce to chefs and sell at the farmers' market on Wednesday and Saturday. Cas or an employee sells at the Austin Farmers' Market. Sunday is more or less a day off.

Adon, Salome, and Gita Vanwoerden selling produce at the Bayou City Farmers' Market.

Although Gita does not enjoy the activity of selling in and of itself, she does enjoy the contact with people and informing them about her vegetables in particular and nutritious eating in general. "I like introducing people to vegetables they've never tried before, maybe haven't even heard of before, like Jerusalem artichokes or Asian long beans, and encouraging them to include these things in their diet," Gita says. "And I really like it when as a result a lot of customers tell me that they do eat more vegetables because of not just me but the other farmers at the market and the chefs and their demonstrations."

What she likes most of all, however, loves, in fact, is farming. "I just love seeing things grow!" she says. "I really, really enjoy planting and taking care of plants."

A few years ago, as their youngest child, Salome, entered her senior year in high school and began applying to college, Gita and Cas sold their house in West University Place and since have been renting a two-bedroom duplex in a much less expensive part of the same general area of Houston. "We had to decide

Main house at Animal Farm.

whether to further develop the farm or continue to maintain a big house for just the two of us," Gita says. "Barely maintain it! Things always need repairing, you know. And besides that, I couldn't keep up with the cleaning, and we couldn't afford someone to come in regularly."

They will always have an apartment in Houston because they enjoy the city, Gita says, and because she does the bulk of her produce marketing there and it remains a business base for Cas as well. But they both are increasingly engaged in their respective ways with

Gita's vegetable growing and marketing and with other projects at the farm.

One of these projects is providing instruction in permaculture, a holistic approach to designing ecologically sound communities and food production systems. In the late seventies, Bill Mollison, an Australian ecologist, coined the term "permaculture" by contracting "permanent" and "agriculture," and his work and writing led to an international movement. Proponents in many countries organize themselves into groups and offer training in permaculture principles and practices, including a series of courses that can lead to certification.

Gita and Cas discovered permaculture during a 1997 vacation to Peru, where they stayed in a lodge on a tributary of the Amazon that was designed according to permaculture principles. Copies of some of Mollison's books were there, and as they read them—with Gita having become a farmer by then and Cas recently completing their off-the-grid farmhouse—they recognized a kindred spirit. When they returned home, they located the nearest permaculture group, which turned out to be in Austin, took the training, and received certification. They then organized a permaculture group in Houston and began offering certification courses themselves at Animal Farm.

Both to develop their design skills and to have adequate physical facilities for accommodating permaculture students, in 1998 Cas and Gita constructed a building they call an education center. They made its walls with straw bales and mud and got a man from South Africa to make a thatched roof with South African grass and reeds. Besides serving the purposes of permaculture training, the center in more recent years has also become the site of a twice-yearly yoga retreat held by their oldest daughter, Dana. A yoga instructor in New York City since graduating from Skidmore College with a degree in dance, Dana brings some of her students to Animal Farm in April and October for four-day retreats. Gita, a yoga practitioner herself when she can get the time, prepares all the food, serving vegetables from her gardens.

Other current or upcoming projects at the farm are various sorts of construction projects. Gita and Cas are in the final stages of completing a guest cottage of straw bales and mud, with interior walls finished in flying concrete. In addition, Cas has a swimming pool under way, one that does not require chlorine for bacterial control but instead has a reverse osmosis filtration system, and Gita is currently building a mud oven near the house. "I love primitive methods of bread baking," she says. Her next project is to make a root cellar. "I need much more storage space and want to depend less on refrigeration," she explains.

They also plan to build another off-the-grid house. The one they live in now is a small two-story structure on a platform elevated eight feet from grade level. The living-dining area on the first story is 20 by 30 feet, and two bedrooms and a bathroom fill the same dimensions on the second story. On the roof, twenty photovoltaic panels provide electricity for lights and small appliances, but not for air conditioning, since many windows, glass sliding doors, and ceiling fans eliminate the need for it. When they need heat in winter, they use a wood-burning stove.

Though Cas loves the house and relishes memories of the four years it took him and their friend to construct it, Gita has never found it comfortable or pleasing to look at, and she talks about it with both amusement and annoyance. "It's a white elephant!" she says, laughing and shaking her head. "It's what happens when you get two guys who were deprived in their childhood of building a treehouse!" Pointing to the winding staircase in the middle of the living room, she says, "See? This staircase is proof!"

Unlike some independent businesspeople, Gita and Cas have never struggled to have enough money. Cas's earnings have always been ample, and, though Gita's farming income for the first few years amounted only to what she describes as pocket money, it has since become a significant supplement to Cas's.

"We've had wonderful animals, and we have this farm, which we've paid off, and we build interesting buildings," Gita says. "And we've sent all three children to college, Dana without scholarships, and Adon and Salome with some, and we've always taken wonderful vacations."

Yet they haven't accumulated any money either. "Neither of us came from families with money, so we haven't inherited money, and we have no savings of any kind," Gita says. "Nor do we have health insurance. We could afford it, but we almost never go to doctors, so we don't want it, don't consider it worth the cost." So far, only one major medical need has arisen—Cas recently had to have a hip replaced—and they located a surgeon in Thailand, where the procedure costs far less than in the United States. "We made a vacation of it," Gita says, "and it worked out fine. Cas did well, we had a good time, and our employees took care of the farm."

Gita says she feels wealthy, and she thinks that Cas and the children do, too. "Not because of what we have," she says, "but because of what we do."

Contact information

Gita and Cas Vanwoerden
Animal Farm
16723 Sycamore Rd.
Cat Spring, TX 78933

Telephone: 979-992-3038
E-mail: gita@trcat.com

Home Sweet Farm

Making American Dreams Come True

Brad and Jenny Stufflebeam's twenty-two-acre farm is in Washington County, a few miles southwest of Brenham and about halfway between Austin and Houston. The land is gently rolling, and the scene is generally pastoral. Most people practicing any kind of agriculture—whether longtime rural residents or city folks with second homes in the area—raise cattle to sell, when the price seems right, to individuals and feedlot buyers who frequent the local auction barns. On almost every farm, you see cattle grazing or big round bales of hay or some of both. But not at Brad and Jenny's. Here, on the twelve acres lying between the road and the creek that divides the farm into almost equal halves, you see three to four acres of noncontiguous vegetable plots, six to seven acres of cover crops, one or two long, narrow greenhouses, big compost piles, and small flocks of ducks, geese, turkeys, and chickens.

What Brad, thirty-six, and Jenny, thirty-five, are doing used to be commonplace, but not any more. To some people, what they are doing does not look like real agriculture at all. This doesn't surprise the couple and ordinarily doesn't give them pause, but in February 2007 they were notified by the Washington County Appraisal District that their farm no longer qualified for an agricultural tax exemption. It did not "meet the degree of intensity requirements" necessary for the agricultural rate and would consequently be taxed in the future at the much higher residential rate. Brad e-mailed the news to their customers and friends, asking for letters of support in the appeal he planned to make. Though anxious, he enjoyed the irony of the term "intensity" and the notion that not enough of it was evident in their farming. "If it gets any more intense here," he wrote in one e-mail, "I don't think I can hold up to it!"

Barely two years since purchasing the farm, in December 2004, and moving with their two young daughters—Carina, born in 1997, and Brooke, born in 1999—from Elm Mott, just outside of Waco, Brad and Jenny were achieving their goal of raising vegetables organically and selling them to customers in their CSA program. Despite a drought during 2005 and 2006, most of the Stufflebeams' vegetables flourished, irrigated by a drip tape system Brad installed to deliver water from their well. And, with only part-time help, Brad and Jenny were able to develop their business quickly. They increased their vegetable cultivation from an initial one-third of an acre to four, and they more than tripled their CSA membership, expanding from twenty-five subscribing households to eighty-five.

The tax appeal, they decided, would be an opportunity to introduce the appraisal board to new models of organic and sustainable family farming, of necessity

Field of greens and Sudan grass at Home Sweet Farm.

focusing on theirs. It would be a chance to discuss the burgeoning local food movement and the economic opportunities it opens to agricultural entrepreneurs. "When people see that farmers can sell food directly to customers and actually get a fair price for their products and make a living," Brad maintains, "then they start thinking outside the box about farming and food." So Brad selected the most pertinent letters from CSA members, and he prepared plenty of spreadsheets: spreadsheets that listed more than a hundred varieties of vegetables and cover crops and their planting dates; spreadsheets that listed a growing customer base and charted a rising income; and spreadsheets comparing his per-acre net income with that of neighboring cattle producers.

Shortly before the April 30, 2007, appeal date, however, *Houston Chronicle* columnist Lisa Gray got wind of the story from some Houston members of Brad and Jenny's CSA, made a quick visit, and wrote it up. Distilling the essence of the Stufflebeams' dilemma and Brad's data, she reached a much broader public than Brad would have been able to, of course, and she also riveted the attention of the Washington County appraisal board and saved Brad the effort of an appeal. The chief appraiser promptly sent two assessors to the farm to walk around and look at things more closely than a single, drive-by appraiser had done before. Close up, and with Brad guiding the tour and explaining what the assessors were seeing, the "degree of intensity requirements" appeared quite sufficient, and Brad and Jenny got their agriculture exemption back.

Educating people about sustainable agriculture is not always convenient, and when something like tax status is at stake it can also be unnerving, but Brad and Jenny consider it to be part and parcel of their farming. It is a theme that comes up repeatedly the cold, drizzly December morning we visit their farm, talking with them as they stoop over rows of arugula, snipping it with orange-handled scissors. After arugula, they work their way to the surrounding rows of turnip greens, baby collards, pak choi, mizuna, kale, swiss chard, spinach, red dandelion greens, mustards, leeks, cilantro, watercress, salad burnet, baby lettuces, and radicchio. In some of the rows, tall stalks of Sudan grass—dead, dry, and golden—droop over the greens, rustling now and then in the raw breeze.

"We had this whole plot in Sudan grass, for a cover crop, but it rained so much this past spring, I couldn't get my little tractor in here to cut it, and it reseeded on me," Brad explains, then chuckles. "So this isn't intentional!"

"No," Jenny says, smiling, "but I think the dry stalks are kind of beautiful. I like how they stand out from all the wet green. I like the contrast of color and texture."

Brad is thin and has a dark complexion, black hair, and a black beard with just a little gray. His fervor for the things he loves most—his family and their work—registers over and over in the fluctuating rhythms and tones of his voice. Jenny is short and has honey-brown hair and a round, tawny face. She is articulate and expressive but not as demonstrably fervid as Brad.

Immediately prior to operating their own farm, they worked and lived for two and a half years at the World Hunger Relief International (WHRI) farm in Elm Mott, a training ground for people going into third-world farming and rural development. There, from 2002 to 2004, Brad managed the farm, including cultivating vegetables organically for a twenty-two-member CSA, running a Grade A goat dairy, and helping the executive director train interns. Brad considered the position not just a job—which, with a wife, a three-year-old, and a five-year-old to support, he needed—but also their family's training to eventually farm and work in rural development somewhere in Central America. He and Jenny had come to share this vision during the course of owning and operating a successful organic nursery and landscaping business in McKinney, just north of Dallas, from 1994, the first full year of marriage, until 2002, when they sold it to go to WHRI.

The transition from commercial horticulture to farming quickly proved to be more exacting even than Brad had expected it would be. It turned out that the executive director of WHRI resigned about the time Brad was hired. "So it fell to me to hold down the whole place, farming and everything, for the first eight months," Brad says. "I actually planned the crops for the next CSA season on my own time, before I was on the payroll, and then once I got there I had to get the seeds going in the greenhouse, run the dairy, sow cover crops, work with four interns, give visitors tours, and make sure that a part-time fundraiser kept some money coming in and got the bills paid." To compound the challenges, much

Jenny and Brad Stufflebeam harvest arugula.

of the equipment needed repair, so Brad repaired tractors and implements and also rebuilt the pipeline that delivered water from the well to the interns' small apartment and the trailer the Stufflebeam family occupied. "Oh man," he says, "what better way to learn how to farm!"

He got all the machines fixed and all the crops planted, took the CSA from twenty-two members to sixty-five within a year, and helped a new executive director get going as well. Learning so much so fast, and training interns who came in knowing even less than he had known, Brad, together with Jenny, soon began to reflect on how little they and the people they were training actually knew about organic agriculture compared with so many of the people in third-world countries they supposedly would be helping. "We started to question how do we really help these people," Brad explains. "So many of them are already farmers, so what do we think we're doing? Because the main method we're taught here in our country is you're supposed to buy things, buy all this equipment, big equipment, big tractors, lots of chemicals, lots of seeds. And then you're in debt and your expensive equipment breaks down, and in third-world countries and here sometimes, too, it won't ever be fixed, or it keeps breaking down and having to be

Jenny Stufflebeam bags arugula for a special order.

Brad, Jenny, Brooke, and Carina Stufflebeam.

fixed, and it all ends up being a terrible investment. It's not economically or environmentally sustainable!"

Even in their nursery business, they had dealt over and over with customers who knew nothing about more sustainable methods of growing things. "People would show up at our nursery in June, wanting tomatoes to plant, and we'd tell them we didn't have tomatoes because it was too late to plant," Jenny says. "And we'd explain about the growing seasons in our area, and that they might find tomato plants at Home Depot and chain nurseries but planting them too late would be a waste of their money and time. Things like that came up all the time, and we just kept seeing how few people here know how to garden, much less farm."

"And so at WHRI," Brad says, "we started realizing that if we were going to farm sustainably and help bring other people into this kind of farming, there's as much need here in the U.S. as in the third world."

Shifting their sights from Central America to Texas, they looked for land while working at WHRI, where their only expenses were their personal telephone and gas for their car. "We saved most of my salary, and with our room and board included in my job, we could take our time locating a farm to buy," Brad says.

"Too much time!" Jenny says, chiding Brad. "He was very picky."

"I was picky," Brad agrees, and explains that he wanted to be east of I-35 because rainfall is greater than west of it, and he wanted to be near enough to at least one city to build an urban customer base. So he honed in on the more or less triangular area formed by Houston, San Antonio, and Austin. He used the *Handbook of Texas* to research the agricultural history of specific counties, soil types, rainfall, and average seasonal temperatures, especially during winter because they wanted to grow things year round. He used the Internet for locating properties up for sale. "And in our time off from WHRI, we'd take the girls and find a campground near things I wanted to look at."

A time or two, they ventured south of their target area and eventually came across the tiny community of Sweethome, a little east of Hallettsville. Though too far from the urban areas they had in mind for their farm, the name "Sweethome" resonated with them. "We played with it and came up with Home Sweet Farm as the name we'd give our farm. It just seemed to sum up what we were looking for, you know, symbolize our dream."

The twenty-two acres they ended up purchasing had been for sale only two weeks when Brad found the listing on the Internet. The location seemed ideal, the price within their budget, and the option of owner financing a great boon. Since Jenny and the girls had declared a moratorium on camping for a while, Brad took the first look by himself. He found a serviceable white frame house, a small metal barn, and a desirable combination of soils—sandy loam and two types of black soils, both of them clays but not too dense. He liked the way Dogwood Creek flows through the middle of the property and also liked the post oaks, native pecans, and hickories growing along its banks. And, as he walked and rewalked the place, he discovered eight old terraces for controlling erosion that were still doing their job, a feature that surprised and delighted him.

"So we jumped on it," Jenny says. "We knew the value was right and we weren't going to play negotiation games. We'd saved money from the sale of our nursery and most of our WHRI pay, so we were ready to go. We made an offer right away."

"And closed twenty days later," Brad adds, "then moved in December 2004 and hit the ground running."

Having managed the CSA of WHRI for more than two years, Brad knew it was feasible to start by cultivating a third of an acre and aiming for twenty-five members, so he began preparing the soil and starting seeds in a 10- by 10-foot PVC-framed portable greenhouse that WHRI let him take on loan (since replaced by the 16- by 96-foot ones Brad builds himself). He also got the farm website up and running. During the months of dreaming and looking for land, he had been creating the site. "Our website was and still is basically both my business plan and my tool for getting customers," Brad explains. "I'd done all the research and knew the links and search engines I needed to turn up on when I was ready to launch. And I also got us on pickyourown.com and localharvest.org, and so, when we were ready, boom, our farm came up."

They also publicized their CSA through Katy area Christian homeschooling associations, which, being homeschoolers themselves, they were already somewhat familiar with. In addition, they interested a Brenham newspaper reporter into writing about their farming, their CSA, and their desire to work with their neighbors to revive local agriculture and create a local food community. "We had no trouble getting twenty-five subscribers," Brad says. "In fact, we had to start a waiting list for those who wanted to get in as we expanded our production and our CSA."

As if farming and developing their first direct retail market were not enough to do, Brad also assumed a leadership role in TOFGA, serving first as a regional representative and then in 2006 becoming president for a two-year term. Though TOFGA is a volunteer organization with a shoestring budget that comes largely from membership dues, its board of officers in recent years has held the largest annual conference on organic agriculture in the state. The conference includes fieldtrips to farms near the conference site and presentations by leading farmers and others involved in organic food production and consumption in Texas and other states. Its main purpose is to educate the public about the benefits of, and opportunities in, organic agriculture and to be a resource for novice farmers and market gardeners. For the same purpose, TOFGA also holds periodic workshops for prospective farmers and market gardeners, providing technical information about production as well as strategies for marketing.

Serving this purpose is what drew Brad into his leadership roles. Not only has he organized conferences and workshops, but he has greatly expanded TOFGA. He created one of the main features of the website, the Texas Local Food Locator, which lists farms by region of Texas, helping farmers and customers find each other. "Jenny and I've always been service oriented, education oriented, and I really want to help grow new growers and nurture local food communities all over the state," Brad explains. "I'm not just growing and selling organic vegetables, you know. I'm growing and selling the local food movement."

He is also working to expand the type of services TOFGA provides beyond the educational and informational. He recently initiated discussions with an insurance company about medical insurance for member farmers, many of whom lack it, including the Stufflebeam family. "It looks promising," he says.

Brad and Jenny met at Plano High School and married at the end of 1993, four years after he graduated and completed U.S. Navy duty in the first Iraq war, and three years after Jenny graduated. As children of divorced parents, Brad and Jenny shared a determination to marry and remain married for life, and to somehow integrate raising a family with making a living. "We were both committed to organizing a life where we, as the dad and mom, would be with our children all the time," Brad says. And though both had to some extent been raised as Roman Catholics, they thought of themselves mainly as nondenominational conservative Christians and shared a desire to develop their own home-centered form of worship and Bible study. "We've always sought wholeness," Jenny says, "not compartmentalization." Their dream of farming and helping others get into farming developed from these shared desires and from the convergence of their individual but complementary interests: Brad's in entrepreneurship, horticulture, and good food, especially Italian, and Jenny's in nutrition.

Brad was born in Richardson in 1971 and grew up in Plano with a sister five years his junior, children of an electrical engineer and a full-time homemaker. Throughout his childhood, visits to the Little Italy neighborhood of Des Moines, Iowa, kept him connected to his extended Italian American family on his mother's side. His great-grandfather owned and operated a corner deli and also grew an enormous vegetable garden. "He bought an empty lot next door to his house just for a garden and an orchard," Brad says, "and he made wine out of all kinds of fruit and berries. Growing and cooking good food was really an important part of their daily lives." Brad's grandfather also gardened, devoting his entire backyard to it. "They were master gardeners,

American Buff geese foraging near the compost pile at Home Sweet Farm.

and totally organic because even if they'd wanted chemicals, they couldn't have afforded them," Brad notes. When his great-grandfather died, Brad got his collection of early Rodale publications on organic farming and gardening.

Brad's mother was a gardener too, but when Brad was ten a car wreck left her blind and brought her gardening to an end. Brad's parents' marriage also soon ended, and, though his father always lived nearby and remained economically and emotionally involved in Brad and his younger sister's lives, Brad assumed many household management and maintenance duties. "My mom was awesome. She did a great job of raising us all the way through," Brad says. "But my childhood ended kind of early. I started paying the bills and balancing the checkbook, keeping the yard, things like that. I grew up fast."

By the time he was twelve, he was running his own neighborhood lawn-mowing business, and within two years he bought two additional lawn mowers, a weed eater, and a blower. "I hired other kids in the neighborhood to help me, and we maintained twenty-two lawns and brought in pretty good money," Brad explains. "Sometimes, though, some of my friends—my employees!—just wouldn't take it seriously," he adds, laughing, "and I'd have to make them go back somewhere and edge again, or prune something they forgot. 'We're not out here to play,' I'd remind them, and I had to fire a few. But mostly it worked fine, and I paid for my first car and have been working ever since."

Jenny was born in 1972 in St. Louis, the middle child of three half siblings—one from her mother's previous marriage and one from her father's subsequent marriage after his marriage to Jenny's mother ended in divorce, when Jenny was two and a half. Jenny lived with her father and stepmother. Her father worked as a regional sales manager for Ralston Purina and was often transferred, moving the family to California, Illinois, Georgia, and a couple of times to Texas, including, toward the end of Jenny's childhood, to Plano.

In high school, Jenny got interested in nutrition and in nursing and sometimes worked as a nurse's aide. After graduating, she attended nursing school at the University of Arkansas for a semester and then studied nursing for a semester at the University of Texas at Arlington. She decided that was enough. "College just wasn't my cup of tea," she says, "but nutrition was, although taking nutrition courses, I realized there was something wrong about what I was learning, a piece of the puzzle missing. In nursing school and in hospitals, I saw curative medicine, and what I was pursuing was alternative medicine, preventative medicine, which is what nutrition is, really." She does not regret attending nursing school and working in hospitals, however. "It formed my decision to study nutrition on my own and try to put into practice what I learn, rather than go to the doctor when I'm sick."

Though Jenny and Brad and their daughters are omnivores of natural foods now, Jenny says that her nutritional research took her through every diet system. "Veganism, raw foods," she says, "you name it!"

"That's when you were taking it too far," Brad chimes in, laughing. "When she got into raw food, I was like, wait a minute! This is taking all the pleasure out of life!"

"Before that," Jenny says, "I took his cheese away and he said no to that, too!"

Being commercial vegetable farmers, their diet is rich in fresh vegetables throughout the year, but they also raise poultry for their own consumption as well as for insect control and soil fertilization in their vegetable plots. They prefer heritage breeds, for the flavor of their meat, for eggs in the case of ducks and chickens, and for traits that lend them to doing the jobs they are intended to do prior to slaughter. Their current flocks include Buff Runner ducks, American Buff geese, Bourbon Red turkeys, and a colorful variety of chicken breeds, including Wyandottes, Barred Rocks, McMurrays, Buff Cornish, Araucanas, and Cuckoo Marans.

They recently purchased two Brown Swiss calves for the girls, on the occasion of Carina's tenth birthday. "We bought the girls a couple of future milk cows," Brad says. "In eighteen months, the girls will be milking cows and have two new calves that in another eighteen months they can sell as beef, if they want, and have money of their own, for another farm enterprise of their own, if that's what they want." The girls love all the animals, he notes, and he and Jenny are teaching them that on a farm all animals ideally serve more than one purpose. "And the purpose of these calves isn't just to become a source of milk and beef," Brad goes on, "but to get the girls to understand the value of hard work. I believe in the American dream. Maybe it's naïve, but I do. I believe that if you have a dream and you're determined and work hard enough, anything is possible. And Jenny and I want our girls to grow up believing it and experiencing it for themselves."

Home Sweet Farm CSA currently includes a hundred subscribing households, four times the number Brad and Jenny began with three years ago. They enroll members on an annual basis and charge $960 for a thirty-two week share (the equivalent of $30 per weekly amount of produce), which may be paid in total or in three installments. They have four "drop sites," locations where weekly shares of produce are available for members to pick up: one in southwest Houston, one in northwest Houston, one in Brenham, and Home Sweet Farm itself. Depending on the location, the produce is delivered either on Tuesday or Thursday afternoon, and though Jenny drove the delivery rounds in the past they recently hired a driver, a retired fireman from Brenham. "He says he likes having extra beer money, and it sure makes our life easier," Brad says. "Especially mine!" Jenny adds.

For the foreseeable future, they plan to hold the CSA to a hundred subscribers and yet expand production. "We want to supply our members with more quantity and better quality," Brad says. They also are selling to more chefs. Monica Pope, owner and chef of t'afia, a noted Houston restaurant, has been one of their main customers all along. To her they have added Bryan Caswell, owner and chef of Reef, another noted Houston restaurant, and also added the café of the Brenham Health Food Store. "One restaurant can equal about ten to twelve CSA memberships," Brad says.

Though their cultivation practices meet or exceed the national organic standards, they have so far chosen not to certify their land and market their produce under the USDA label. This is partly because of the costs and paperwork involved, but also because they sell directly to people who know them and in many cases visit the farm throughout the year. Their customers are welcome to visit and see for themselves and hear from Brad and Jenny what farming methods the couple use. "We get a few hundred visits a year from customer-inspectors," Brad says. "This is the ultimate accountability, for customers to know their farmers and develop a relationship with farmers and food they can trust."

During their first year at Home Sweet Farm, Brad and Jenny did, however, participate in an alternative certification program called Certified Naturally Grown, run by Naturally Grown, Inc., a not-for-profit organization based in Stone Ridge, New York. Administered by an advisory board of farmers from around the country and open to farmers in every state, the program is based on the national organic standards but requires less money and paperwork than certification through a USDA-accredited certifier. It is intended for small family farms with local markets and depends upon participating farmers to inspect the farms of other participants in their area, as called upon by the farmers themselves. This all-volunteer structure has the beginnings of a participatory guarantee system, according to Elizabeth Henderson, a nationally known New York state CSA farmer and author, yet it may not be rigorous enough to fulfill its promise. Brad agrees. "It's a good idea, but it's almost totally dependent on volunteers and an honor system that might or might not work."

Mixed flock of chickens and geese at Home Sweet Farm.

Brad and Jenny, in addition to running their CSA, hold almost monthly market days at their farm, as a function of a local organic food cooperative they organized during the past year. Other family farmers join the Stufflebeams and make a wide range of products available for sale on a Sunday afternoon, including meat, eggs, cheese, honey, fruit, and vegetables. Ironically, Brad and Jenny have so far been unable to provide vegetables for the market days because all their supply has gone to their CSA. With expanded production, however, this will no longer be the case.

Brad publicizes the market days in the local newspapers,

through their website, and by e-mail to their CSA, and he and Jenny operate it on a membership basis, charging all customers except those already in their CSA a small annual fee. The customers are mostly people who live in the outer suburbs of Houston, plus Houston and Austin people who live full time or part time in rural Brenham, and the venue is proving increasingly successful for all the farmers involved. Brad and Jenny hope that eventually the clientele will also include locals who don't hail from cities. "Some of our neighbors show up and look around," Jenny says, "but they don't buy. They're just curious."

Brad attributes their reticence mostly to the prevalent expectation that food should be cheap. "If they could only see that cheap food is why there's mostly huge corporate farmers instead of farmers like us," Brad says, smiling and shaking his head. "Farmers like they could be, too, if they could just see beyond cheap!"

He is not giving up on them, though, and he and Jenny are about to make a move that will call more local attention to their kind of farming and marketing than ever before. They have signed a lease, effective January 2008, for a 112-acre farm five miles up the road. Its owner is from San Antonio, for health reasons cannot spend much time on the farm, but wants to keep the land in agricultural production. Observing what Brad and Jenny are doing, she began considering possibilities other than cattle and hay, talked with them, and they worked out a deal. Her farm is one of the oldest in the area and has six barns and a hundred-year-old, fully restored house, which Brad, Jenny, and the girls will live in. "We'll be restoring the soil, and we'll plant some major crops of potatoes, sweet potatoes, onions, watermelons and other things that we can't put in enough of here on our smaller farm," Brad says. "Plus, this bigger farm will give us other opportunities for expanding production, like doing grass-fed beef and lamb."

They will keep Home Sweet Farm and continue to grow vegetables here, but with help. A couple they became friends with through WHRI will live on the farm, in the house the Stufflebeams now occupy, and manage production here and also help manage the CSA. Brad, ever the public educator, hopes that both aspects of this new arrangement—leasing the larger farm and making available the smaller farm to their co-farmers—will serve as a model. "With the price of land going up and up, I think more and more farmers are going to have to share land and lease land," Brad says, "and I'd like to turn the lady who owns this 112 acres into a champion, an example of what other landowners can do. They can keep their agricultural exemption and help grow the local food community at the same time!"

Excited though they are about the changes in the offing, Brad and Jenny find themselves already feeling a bit nostalgic for the twenty-two acres they are about to leave and the family life they have known here, the pleasures along with the hardships. What they have most enjoyed is simply working together as a family and growing good food. "Making it a living," Brad emphasizes. What they have enjoyed least—and consider the only real hardship—is hiring and keeping help on terms they can afford. For a little over a year they employed a young, single man. He worked half time for them and half time for a neighbor with a small dairy operation, both goat and cow, which seemed ideal all around, since neither the Stufflebeams nor the dairy woman could afford the full-time work the man needed. But, as young people often do, the helper reached the point that he wanted to move on to another area and another kind of work. "And wouldn't you know it?" Brad says. "He quit right in the middle of our fall planting! Which put everything totally back on me and Jenny, of course."

Having co-farmers in the near future promises to solve the labor problem. And yet, left in the lurch, Brad and Jenny discovered that, though they gain something by having help, they lose something, too. "A few days after the guy left, and Jenny and I were out here in our vegetable patch doing everything ourselves, she said, 'You know, I've really missed this, just working by ourselves, and getting to talk.' And I feel the same way," Brad says. "We both just really love working together, and talking and dreaming together." He exchanges a smile with Jenny, then looks down, lifts a foot off the wet ground, and adds, "Not to mention scraping mud off each other's boots!"

Contact information

Home Sweet Farm
7800 FM 2502
Brenham, TX 77833

Telephone: 979-251-9922
E-mail: info@homesweetfarm.com
Web: www.homesweetfarm.com

Shrimp and Meat

Permian Sea Organics shrimp pond at dawn.

Permian Sea Organics

Aquaculture in the Desert

Bart Reid grows shrimp in the desert. In Imperial, Texas, to be exact: a Trans-Pecos town of three hundred people at the junction of farm-to-market roads 11 and 1053, fifty-five miles southwest of Odessa and thirty miles north-northeast of Ft. Stockton. The sky is big here, and the land is flat and dotted with mesquite, tumbleweed, prickly pear, and pump jacks. Named for the ancient ocean that once covered the area and left it rich in oil, Permian Sea Shrimp Farm is not a patchwork of fields but a grid of levees and sixteen ponds. Each pond is four feet deep and four acres across and capable of holding sixteen acre-feet of water, about five million gallons. Since the levies are fringed with vegetation and more or less level with the surrounding land, the grid does not stand out from the road. From there, you see only hints of it, and only if you are trying to make it out, somewhere beyond the brown frame farmhouse fronting the property and facing the road. To view the watery grid, you have to drive or walk along the levees.

The water comes from the Pecos alluvium, an aquifer tapped at depths of 30 to 200 feet by Bart's three wells. It is as salty as bay water, about twelve parts per thousand, and consequently useless for people and livestock to drink, or for irrigating crops. But it is perfect for raising Pacific white shrimp, which Bart, a lean man with dark brown eyes and crew-cut hair, has been doing since 1992, stocking a varying number of ponds in May and harvesting during several weekends from late October into November. And the aquifer, with hundreds of millions of acre-feet of largely untapped water, is not being depleted. "All the water we bring up," Bart says, "we put in a pond and, outside of what evaporates, it's going to percolate back into the ground. We're not using up what we pump out. The way most fields of row crops are irrigated, only a little bit of water goes down into the soil, but most of it goes up, gets sprayed into the air. In a pond, though, the water's confined and there's less evaporation per volume of water than in row crop operations. Sure, you've got a layer on top that's evaporating, and in the worst conditions, say forty-mile-per-hour winds and 110 degree heat, up to two feet per day can evaporate. But that's not most of the time, and what doesn't evaporate is going to end up back in the aquifer, because we reuse our water."

Shrimp is the most popular seafood in the United States, and as overharvesting has depleted wild populations of shrimp and many other aquatic taxa, aquaculture has increased in the United States and throughout the world. Along with addressing the problems of depletion, aquaculture presents a way to produce seafood that is not contaminated with toxins such as mercury and PCBs (polychlorinated biphenyls), industrial and agricultural chemicals that are the main reason health warnings advise limited consumption of wild seafood especially prone to high concentrations. But conventional aquatic farming, both inland and coastal, relies on methods that damage the environment and pose some of the same risks to human health that wild-caught seafood does.

Much as conventional dirt farmers regard the soil as nothing but a lifeless medium for delivering chemicals to plants, so conventional fish farmers regard water. They stock their pools densely, treat the water with hormones, antibiotics, chemical fertilizers and pesticides, and feed the fish mainly ground-up wild fish, or fishmeal, which only concentrates the levels of mercury, PCBs, and other toxins found in wild seafood. These chemical cesspools, as they are sometimes described, in coastal areas contribute to the contamination and depletion of wild fisheries and also endanger or destroy aquatic and coastal plant communities. Inland, conventional aquaculture often wastes potable groundwater and also pollutes it.

"Conventional aquaculture is the equivalent of putting cattle in a feedlot instead of pasturing them," Bart explains. "In an organic system, we don't use chemicals and we don't stock so densely that the pond can't absorb the fish waste and convert it to fertilizer. We create and maintain a healthy pond environment so the fish produce their own natural flora to live on."

Bart, born in Denton in 1963, is not only a farmer but a marine biologist and with his wife Patsy, born in Ft. Worth in 1965, also a restaurateur. They own, manage, and, assisted by only one part-time employee, do most of the cooking at their restaurant, the Permian Sea Shrimp Store. In addition to shrimp, sauteed or fried, the menu includes Patsy's homemade coleslaw, beans, and pies: pecan joy, gooey butter, peanut butter polly, and Hershey Bar pie. Bart describes the store as their farm stand, and it's a popular one, despite its remoteness from towns of much size. People drive from miles around for lunch on Monday through Saturday and for supper on Friday night. They can also purchase frozen shrimp to take with them. A tan, steel building with wide windows, the store is located almost at the main crossroads of town, about a quarter mile from the farm, where Bart and Patsy live with their son Colton, born in 1991, and two daughters, Ryan, born in 1994, and Hannah in 1996. "Our store's an easy walk from the farmhouse for the kids," Patsy says. And in the course of our two-day visit for harvest—a crisp, clear Friday and Saturday in late October—we go back and forth between the farm and store many times ourselves, talking with all the Reids and also Patsy's parents, visiting from Ft. Worth, and with friends, neighbors, and customers, including a van full from Artesia, New Mexico, about three and a half hours away. A bright yellow flag mounted on an entry gatepost at the farm reads "Fresh Harvested Shrimp Today," welcoming returning customers and first-timers alike.

Bart schedules the harvest each fall according to a succession of cold fronts. "As the fronts come in and the pond water cools, shrimp slow in their growth and gradually stop," Bart explains. "Harvest is when they won't grow any more." Given the vagaries of weather, the date of the first weekend in the harvest series cannot be determined very far in advance, and beginning in early October Bart and Patsy get an increasing number of phone calls each day from customers who want to keep posted. Some simply place orders for a certain quantity of shrimp to pick up at the store on one of the harvest days, but many plan to show up at the levees as early on a harvest morning as possible. They like seeing

Bart Reid readies the shrimp pump for the fall harvest.

the shrimp pumped out of the water and into vats of ice for "chill killing," then fill up their ice chests and head home. Others, mainly friends and neighbors, plan to help retrieve by hand the shrimp stranded on the pond bottom after the pumping is done. Although Bart grows enough shrimp to supply the Permian Sea Shrimp Store throughout the year and to sell wholesale to several grocery stores and restaurants in Alpine and Marathon, the retail sales on harvest days are, dollar for dollar, the most profitable. Shrimp sold fresh from the ponds are shrimp that do not have to be trucked to a processor on the Gulf coast for cleaning and freezing and then trucked back. So Bart and Patsy welcome all comers to the levees on harvest weekends, even if the many phone calls in advance make both running the restaurant and preparing for harvest more hectic than usual.

Bart and Patsy are used to dealing with a wide range of people on a daily basis—oilfield workers and farmers, bankers and preachers and teachers, and travelers of all kinds, from motorcyclists to RVers and truckers, checking out the territory between I-10, to Imperial's south, and I-20 to its north. Although both converse readily and candidly, Bart tends to be sharp and droll and Patsy effusive and warm, complementing each other like salt and

Bart Reid and helpers take shrimp from the separator to be chill-killed.

honey. The restaurant, which opened in January 2002, was Bart's idea, and Patsy went along. The last of eleven children in a Catholic family, she was taught that once she married she should defer to her husband's wishes and decisions, a precept and practice that she continues to embrace. But however deferential she may be to Bart, she does not always agree with him, and she was not enthusiastic about the restaurant. "I didn't want to get broad as the side of a barn or breathe grease all day and smell like grease," she remembers. "But you know what? You just have to say, okay, grease smells like money! And I love people and enjoy all the people who come eat here and inquire about us and our store and farm." Patsy has fair hair and skin, big blue eyes, and deep dimples in her cheeks whenever she smiles and laughs, which she does almost constantly. "I'm a bigger ham than I ever dreamed I was!" she admits. "And honestly, what I enjoy most about our whole business is seeing Bart out there working on the farm, keeping those ponds going. It's like Bart is the farmer and the farmer is Bart. The two are synonymous, and the joy on his face makes me very happy, makes everything we do worthwhile."

Like many farmers, Bart is something of an engineer, and he designed not only his ponds but also his harvest equipment. "The ponds are built with earthmovers that have lasers that grade the pond bottoms and levee tops exactly so that when you drain the water—shrimp have to come out with the water—you won't have humps in your pump line, you don't have air pockets, and the shrimp come right out," he says, then grins wryly and chuckles. "Most of them, anyway. The pond I'm about to harvest now has all this seaweed, and pumping's not going to work like it does when the bottoms are cleaner." The ponds are also equipped with electric aerators—small floating platforms that carry four windmill-like turbines—but Bart now maintains such low stocking densities that he no longer uses them. Harvesting equipment consists of a yellow steel pump fueled by a diesel generator; blue and white vinyl hoses; and a blue shrimp box, a device which, as pond water is pumped into it, traps the shrimp and allows the water to flow on, into hoses that take it to other ponds. Though Bart does almost all of the farm work himself, he employs a man to help him part time, mainly from summer through harvest. "As we harvest a pond, we move its water into another pond," he explains, "to save as much water as we can, so we don't have to pump it again next year. If I can save a pond full of water, that's a pond I don't have to run a water well to fill up. Just conservation of the resource."

One of these ponds is used not for raising shrimp but for an endangered fish native to the Pecos River, the Pecos pupfish. Protecting this species is a joint project of Texas Parks and Wildlife and the U.S. Fish and Wildlife Service, and Bart has a contract with these agencies to provide habitat and safe harbor for the fish. Though the water in Bart's ponds is saltier than the naturally salty Pecos River, it is not too salty for the pupfish. He is also creating a wetland habitat for the threatened puzzle sunflower, also called the Pecos sunflower. Its natural habitat is the Diamond Y Preserve, an endangered desert marshland managed by the Nature Conservancy fifteen miles south-southeast of Bart's farm. Bart does these projects under the auspices of the Organic Aquaculture Institute (OAI), a nonprofit research organization he founded and directs. The OAI brings together leading academic researchers, entrepreneurs, and aquaculturists to research, develop, and promote the organic production of aquatic food animals as well as to propagate and protect endangered local species and their environments.

Another current OAI project, funded by commercial producers of organic feed, is Bart's use and evaluation of various rations of organic soy and flax in raising shrimp. "This is probably the first year anybody's ever not fed shrimp fishmeal. Most university people say shrimp won't live without fishmeal. Well, what we do in the OAI is push the edge of that kind of thing, because universities aren't and somebody's got to. So we do real-world stuff instead of doing things in an aquarium and telling everybody that if they extrapolate up to four acres, this is what they'll get," Bart explains. "We do whatever we do in four-acre ponds, and we can see exactly what we get and what works and doesn't work. And not feeding fishmeal, I've done something all my old professors and friends and aquatic nutrition people say you just can't do, you know. I love that!"

In the wild, marine animals eat other marine animals. "That's the one weirdness in the marine world, and the reason why fishmeal is a main feed in conventional fish farming," Bart elaborates. "And some European countries allow fishmeal in organic aquaculture, but the bottom line is that the wild fish and the fisheries are not organic systems, and so you shouldn't use fishmeal from those fish as an input in organics. If there was a source of organic fishmeal from an organic fish farm, that'd be one thing, but there's not. I'm committed to the highest organic standards and trying to stay ahead of the curve, so we're not using any fishmeal in our feed."

Issues about feed—what should and should not be allowed for certification in the National Organic Program (NOP)—are as contentious in aquaculture as in beef, poultry, pork, and lamb production. With the implementation of the NOP in the fall of 2002, the USDA allowed aquafarmers to qualify for organic certification by following livestock production rules, which Bart had been doing since 1997, and Permian Sea Shrimp Farm became the first seafood facility to be certified, in 2003. But in April 2004, USDA secretary Ann Veneman rescinded this policy, effectively removing aquaculture from the NOP for an indefinite period, and directed the National Organic Standards Board (NOSB) and an aquaculture task force to devise standards independent of those for livestock. Bart serves on this task force, which was charged with proposing standards not just for feed but for all aspects of organic aquaculture, including habitat, water quality, stocking densities, and treatment of animals. The task force submitted its proposal to the NOSB in January 2006. The proposed rules are now subject to public comment and possible revision, with adoption into the NOP not expected for several years, perhaps as late as 2012.

The abrupt, unanticipated removal of aquaculture from the NOP in 2004 nearly ruined Bart and Patsy financially. Having achieved organic certification the year before through Quality Certification Services (QCS), a well-respected Florida-based certifier accredited by the USDA, Bart was keeping production high, stocking most of his sixteen ponds. He harvested more than 100,000 pounds of shrimp in the fall of 2003, providing the large inventory he needed to sell during the coming year at premium, organic prices. But the USDA blindsided him and forced him back into a wholesale market he thought he was finished with—the global, conventional market that is driven down by shrimp exported from Asia, where production costs are much lower than in the United States and many other nations. "That USDA decision cost us a fortune, because we had all this investment in organic product. We had the product. We had people—big wholesalers—ready to buy. We even had it on the way to some of our customers! And then when the USDA rescinded certification, these buyers wouldn't take it because it wouldn't have the USDA organic label. So we had to sell most of our shrimp for whatever we could get, just to pay bills," Bart explains, his tone exasperated. "It was all we could do to keep from completely going out. I mean, they ruined our business! They left us in the dust!"

Since this nearly fatal setback, Bart has lowered his volume of production, stocking four ponds in 2005, for example, for a maximum harvest of 6,000 pounds of shrimp per pond, or 24,000 pounds total, one-fourth his previous peak volume for the certified organic market. Supplemented by frozen inventory from previous harvests, 24,000 pounds is enough for the Permian Sea Shrimp Store, local retail sales, and wholesale accounts with local groceries and restaurants. If local demand grows, as rising oil patch profits indicate that it is, Bart plans to put from two to six more ponds in production for larger harvests in the near future. And to further relieve some of the pressure on the shrimp supply, he and Patsy are beginning to purchase farm-raised catfish to put on the restaurant menu. "I'm happy at this point keeping things pretty scaled down," Bart says. "There was a time when I wouldn't have been, but we got shafted, you know. We almost got our heads cut off. So I'm taking care of what we have." Their restaurant, he emphasizes, has proved to be the key to sustaining them in shrimp farming as a livelihood. "Without it, we'd have gone under for sure."

Undeterred, Bart and Patsy are more eager than ever for organic aquaculture standards to be resolved and implemented. "Once we can put a USDA symbol on our organic shrimp, then we might scale up and look at the Whole Foods of the world or the Wild Oats and the bigger places," Bart speculates. "And we might

Customers wait while Patsy Reid and a helper dehead shrimp.

eventually diversify what we grow, raise other saltwater fish, red fish and snapper and ling." Patsy adds, "The whole organic thing is taking way longer than we'd like, but honestly, both Bart and I see our future in organic seafood."

She notes, however, that organic farming in and of itself is not for her the be-all and end-all that it and other environmental practices seem to be for some. "I worship the creator," she emphasizes, "not the creation." Bart shares this attitude. "I don't worship the

Patsy Reid fries shrimp at the Permian Sea Shrimp Store.

organic gods, either," he says. "But on the other hand I don't just do organics because I think there's a market and I can make some money. I believe we can do better in agriculture, and grow things the right way, and not have to eat crap from everywhere else that's grown with chemicals banned in this country yet used on crops in other places and sold right back to us. I like the whole organic movement and, even though there are now some humongous organic producers, which some people in the organic world didn't want to see happen, I think organics still benefits small farmers."

Although for now he can't use the USDA organic label, Bart is maintaining his organic certification through QCS, which he admires for being as stringent as ever. "We can't use chemicals or preservatives, and we have to consider animal welfare, which means lower stocking densities and more time fertilizing the water, trying to make an ecosystem of the pond as opposed to just feeding fish," he elaborates. "We fertilize our ponds with composted dairy waste, manure, and silage. It's composted at the dairy and we bring it out and fertilize the ponds with that. With lower stocking densities, I don't need electrical aeration, which lowers my operating costs and is also just part of creating a more natural, more self-sustaining ecosystem. The ponds have plankton, critters that will live in a pond, so you stimulate all that and provide a natural source of food for shrimp as opposed to having a sterile pond and giving them everything. And we can label our product organic, just not USDA organic."

But until USDA establishes organic aquaculture rules, other seafood farmers, who may or may not adhere to such stringent standards, are also labeling their products simply "organic." This leaves consumers in the dark about what in fact has gone into the "organic" seafood they are buying, similar to the way labeling meat "natural" does—short of investigating each producer, anyway—and gives Bart little if any market advantage over conventional seafood farmers. As the long, arduous process of resolving USDA rules for organic aquaculture continues, however, Bart keeps his worst worries and deepest wishes in check. "Doesn't pay to worry," he says, "just like it doesn't pay to worry going into a harvest. Worrying won't make any more shrimp in those ponds if they're not there. But at the same time, I'm trying to fight the old farmer mentality of optimism, too. I mean, optimism is what's great about farmers, but that's also their downfall—thinking, like, we're really going to get 'em next year. Too much optimism and you do stupid things, expecting certain things to definitely happen."

The turbulence of organic marketing and resulting financial hardships—the inability to consistently pay for health insurance, for example—have made Bart more keenly aware than ever that there are easier ways to make a living, but he never imagines getting out of aquaculture. "This is what I always wanted to do. I tend to be one of those people, and it may be a fault, who tries, who never gives up, you know. One of those people who tell themselves, oh, you've got to keep trying. The guy that gives up never reaps any rewards," he says. "But then on the other hand," he goes on, laughing, "you wonder at what point does something become insane. Sometimes you have to quit, right? So I don't know where I am in that deal, but I certainly haven't quit so I'm pushing the insane side for sure!"

Bart earned a bachelor's degree in marine biology from Texas A&M at Galveston in 1986 and a master's from the University of Texas Marine Science Center in Pt. Aransas in 1989. But his fascination with water and the creatures that live in it goes back to his early boyhood. He was raised in Hurst, between Dallas and Ft. Worth, with an older sister and younger brother. His mother was a homemaker, and his father, whose college degree was in biology, managed a sporting goods store and enjoyed outdoor activities, often on his parents' ranch near the Red River. "My grandparents had stock tanks, ponds, and they stocked the ponds with catfish and bass, and while everybody else was messing with cows I was usually down at those ponds. I found them a lot more fascinating than I did cows. So I guess that's where it all started, gigging frogs and catching fish."

Bart and Patsy met in the summer of 1983, when one of Patsy's brothers lived across the street from Bart's parents and Patsy took care of her nieces and nephews. Bart proposed to her in Galveston in the fall of 1985, after an Italian dinner and a bottle of wine. "He walked me out into the surf," Patsy recounts. "A little tropical storm was blowing up, and I had on a dry-clean-only skirt, of all things, and I could tell by how nervous and frightened he was what he was going to do. I was so ready, I wasn't nervous a bit. I was like hurry up!" They married in 1986, during spring break of Bart's senior year in college. Patsy did not go to college after high school but instead worked in retail sales, which she continued to do until Bart finished graduate school. "I like to say I got a PHT," Patsy says, "a putting-hubby-through degree, you know." Bart's first job in commercial aquaculture took them to a redfish farm in Mexico Beach, Florida, where they lived until shortly after Colton, their first child, was born. "With all our family back in Texas, we were feeling the need to come home," Bart explains. "North Florida is beautiful, but Florida's expensive and it rained eighty inches a year, and much as I love the water, I mean, well, I think Texans just want to come back to Texas," he chuckles. "I mean it just happens."

"I kissed Texas ground when we got back," Patsy remembers. "Going across the Sabine from Louisiana into Texas, I said to Bart, 'You pull over because I'm kissing Texas ground!' And he did, and I got out and yeah, buddy, I kissed Texas ground!"

They initially had their doubts, however, about living in the Trans-Pecos. Patsy had never been to this part of the state, and Bart had been here very little. "I'd been out here when I was in college and in grad school. I used to come out here with some mammalogy people, bat people, and catch bats and do crazy things in the mountains," he says. But about the time their first child was born, Texas A&M had begun a shrimp aquaculture research project in Imperial. Bart knew some of the people involved and had heard enough from them to be intrigued. "I came out for a while to watch what they were doing, and it looked promising," Bart says. A man he had worked with in Florida became interested, too, and told Bart he would finance a shrimp farm if Bart would run it. "So in 1992, we moved, and I built the first commercial shrimp farm out here for someone else," Bart says, "and ran it till '93." Then Bart built another farm for another man and ran it until 1997, when he was ready to build his own farm and go into business for himself. "I'd had enough experience to have a lot of ideas about how to improve pond design and ecology and shrimp production, and I also really wanted to pursue organics," he explains, "because for one thing, I was pretty much doing organic already and not getting credit for it or the premium and, second, I could see there was a trend."

For income while building his farm, which took from 1997 into 1998, Bart got an emergency teaching certificate and taught life science and physics to Imperial school students in grades six through twelve. "The school doesn't have enough money to pay me to do that again," he says, laughing. "No, actually," he goes on, "I told the school that if we ever can't get a teacher, I'd try to fill in again. My students did really well, even the ones I thought would never pass anything." Every day after school and on weekends he worked on the farm, constructing ponds and levees, and harvested his first Permian Sea Shrimp crop in 1999.

The years in Imperial have made Bart feel at home both in his work and in the community. "I was always interested in the ocean," he reflects. "I loved everything ocean, but studying marine biology, I realized I'd have to make a living. And as I went through school, I was always drawn to shrimp and fish aquaculture, and it became obvious to me that okay, here's a way to be a marine biologist and possibly make a living other than being an academic. Nothing wrong with being an academic, but I was looking for a more practical real-world way to be a marine biologist. And commercial aquaculture out here in the desert is even more challenging than in coastal areas because you're taking a marine animal and putting him in the desert and creating a marine ecosystem for him to be able to live in. In your ponds, you have to create a whole world that doesn't normally exist here."

The most satisfying moments of his work come from seeing shrimp grow. "I like seeing that they're working,"

Hannah Reid and neighbors harvest the remaining shrimp from an emptied pond.

he says. "I like looking and telling there's a lot of shrimp in a pond." Least satisfying is constantly having to juggle so many things at once, raising three children, running the restaurant, keeping equipment there and at the farm in working order, all on top of growing shrimp and struggling to market it. "And it's all made harder," Bart emphasizes, "by not having enough money to hire nearly enough help."

As novel for the area as shrimp farming is, and all the more so for being organic, Bart doubts that he himself is perceived in the community as being odder than anyone else. "People here are pretty much like people in any small town. Everybody likes to talk about everybody else, but for the most part people accept each other pretty well," he finds, "because everybody here's got some strangeness or why the heck would we be here? I mean, this is a harsh environment. It's a desert. If you were normal, you'd be living in a town somewhere. Everybody out here's got some wart or quirk so nobody really cares, and I think I'm for the most part a regular guy." Besides, Bart notes, their farm and restaurant add to the local economy and give the town a distinction that people seem to like.

He and Patsy contribute to the community in other ways, too. Bart serves on the boards of the water and tax appraisal districts, Patsy serves on the school board, and they and their children are active in the Baptist church. Patsy feels as much at home here as Bart does. "It's a great community. I mean, everything that happens in a big town happens here, and you see it," she says. "And I don't have a problem with my kids going to public school here, the way you do in some places. It's pre-kindergarten through twelfth and there might be 128 kids, so classrooms average about eight to twelve students, and they're expected to study and to behave themselves. I mean, the kids all say yes ma'am and no ma'am and yes sir and no sir, and every parent knows every other parent. Nothing goes on in Imperial without somebody knowing." She laughs, then adds, "Which sometimes is a good thing, and sometimes a bad thing. Anytime you don't want people to know something, you just keep your mouth shut."

Imperial is the only home that Colton, Ryan, and Hannah have ever known. Annual shrimp harvests, as much as their birthdays, mark the passage of time and the course of their growth and development. "We have dozens of pictures of all three of our kids from infancy on playing in pond mud at harvest," Patsy says. The first time Bart harvested shrimp in Imperial, Colton was a toddler and could walk just well enough to follow Bart around on the levees and into drained ponds. During another harvest, Ryan was just old enough to have a head of long hair, and she dipped her long locks in the mud and flipped it off. And Hannah, the youngest, literally cut her first teeth on live shrimp. "She was just barely sitting up, and we sat her in a shrimp basket so she couldn't get out of it or anything," Patsy remembers, "and we just put some of the shrimp out on top of her and they were cold and she picked them up and gnawed on them! They were just the right texture, I guess, and felt good on the gums."

All three children, as far as Bart and Patsy can tell, are happy. They do well in school. Colton, fourteen at the time of our visit, likes science and engineering; Ryan, eleven, likes writing and learning a variety of computer applications; and nine-year-old Hannah likes science and animals. All enjoy outdoor activities with Bart, hunting quail, doves, and ducks, especially, but also watching such nongame birds as the American avocets and black-necked stilts that nest on the farm. If they sometimes wish to live more like their many Ft. Worth and Dallas cousins, it doesn't last. Skateboarding, for example, was such fun one time in Ft. Worth that Colton insisted on getting his own at home, but he soon abandoned it.

"They have a wonderful freedom out here and can do way more interesting things than that," Patsy says, "and I think coming back from their cousins, they realize that. Besides, every summer their cousins beg to come here."

Given a life so filled with family, friends, neighbors, and work, it is little wonder that Patsy and Bart only rarely find time to sit down with just each other. But it does happen. Sometimes, an hour or so before daybreak, when Bart is making the rounds in his truck to check on the ponds, and the children are still asleep, Patsy makes a thermos of coffee and meets him on one of the levees, and they sit and talk in his cab. Other times, they visit late at night. "We love to look at the stars and watch the moon rise," Patsy says, "especially in the summer. The sky out here is really pretty. And hot as the summers get, the wind blowing off the ponds is cool and feels really good. So we sit in our backyard and drink wine and watch whatever's going on in the sky, and talk and just listen to the wind blow the water in the ponds."

These breezy summer nights are among the very best times of all. "Amazing and restful both," Patsy reflects, "like being at the ocean. Honestly, the water in the ponds sounds just like the ocean."

The Reids closed their restaurant in late 2008. Bart's work through OAI has developed enough that they do not need the restaurant for income. Bart continues to stock four to six ponds, mainly with shrimp but also with some flounder and redfish. They sell their shrimp and fish to area restaurants and grocery stores and to individuals who come to the farm. Bart still plans to expand production when the national organic standards for aquaculture are established.

Contact information

Bart and Patsy Reid
Permian Sea Organics
901 East FM 11
Imperial, TX 79743

Telephone: 432-536-2216 and 432-536-2442
E-mail: reid_bart@yahoo.com

Rehoboth Ranch and Windy Meadows Farm

For the Love of God and Family

R**obert and Nancy Hutchins** of Rehoboth Ranch and Mike and Connie Hale of Windy Meadows Farm are friends and business associates with many things in common. Both couples are in their early fifties and live in the gently rolling land outside of Greenville, the Hutchinses on three hundred acres about fifteen miles to the west-northwest of town, the Hales on forty acres about twenty miles to the east-northeast. Both earn their livelihood by raising several species of livestock on organic pastures and selling meat and dairy products to customers in the Dallas area, about an hour's drive to the southwest. The Hutchinses became full-time farmers in 2000, the Hales in 2003, when Robert and Mike left nonagricultural careers spanning about twenty-five years. Both couples have large families. Robert and Nancy are the parents of twelve children, born between 1978 and 1999, all homeschooled. Mike and Connie are the parents of eight children, born between 1982 and 1996, all homeschooled. Both families are also home-churched and describe themselves as fundamentalist Christians. Robert and Mike are among the men who serve as elders in a twenty-five-family congregation that meets every Sunday in the Hutchinses' house.

Neither couple began marriage anticipating raising such a large family, becoming fundamentalist Christians, or farming for a living. It was questioning education systems and deciding on homeschooling that set both couples on similar paths. Mike and Connie made this decision in 1984, when their oldest child was only two. Mike, a math and science teacher in public middle schools before becoming a full-time farmer, read a national report by another public school teacher that argued the superiority of homeschooling to other systems in developing children's intellectual and moral potential. Both he and Connie, who taught earth science in private and public high schools the first four years of marriage, found the report persuasive. And though the Hales were active Presbyterians, the report prompted them to rule out parochial schools as well as public and move wholeheartedly ahead with educating all their children at home.

Robert and Nancy decided on homeschooling in 1985, as their oldest child completed kindergarten at an interdenominational Christian school. They questioned the quality of their son's character development more than anything else. He showed far less interest in studying and learning to cooperate with other children than in competing with them and winning more awards than anyone else. Robert, a U.S. Naval Academy alumnus and high-level defense-industry employee before he began farming, and Nancy, with a history degree from what is now Texas A&M University–Commerce, had been raised

Hale and Hutchins daughters break for a picnic lunch at Windy Meadows Farm.

as Methodists and at the time still were. It seemed to them that, if even a Christian school could not foster the sort of character they wanted, then homeschooling might prove the best means to build character and nurture intellectual growth.

Homeschooling proved to be only the first big step into an alternative way of life. "Homeschooling really opened our eyes to not doing what everyone else is doing," Robert explains during the July day we visited both Rehoboth and Windy Meadows, "and people who start questioning the conventional wisdom about one thing usually go on and question it about any number of others." In his and Nancy's case as well as Mike and Connie's, this meant questioning conventional nutrition and food production and the institutional church. They did not initially realize it, but their questions would lead them, over the next fifteen to twenty years, to radically reshape their lives around not just rearing their children but creating a family business in order to work with them, too.

You don't need to know these facts about their lives, however, to guess that the families who live at Rehoboth and Windy Meadows think differently from a lot of people, differently, for sure, than conventional producers of meat. Looking at their pastures tells you that. Instead of a single animal species typical of conventional livestock operations, there is a variety: chickens, cattle, sheep, and, at Rehoboth, also goats and pigs, with turkeys added to the mix at both places in time to mature for Thanksgiving. Instead of large pastures grazed continuously for long periods of time, there are small segments of pastures, or paddocks, created with temporary fencing. Grouped by species, the animals graze a paddock briefly and then rotate to a new one, leaving just-grazed areas to grow before being grazed again. The configuration of paddocks and animals you see one week may not be what you see the next week.

But of all the things that stand out as different, probably the most riveting is the sight of chickens on pasture—flocks of them in various stages of development—eating grass and insects. In conventional poultry production, chickens are nowhere to be seen unless you go into the enormous steel buildings confining them, some as huge as 40 feet wide, 8 feet high, and 470 feet long. And since the national organic rules require only that animals have "access to the outdoors" and not that they actually spend a certain amount of time on pasture, confinement in large buildings is also the norm for industrial organic operations. Such buildings are conspicuous by their absence at Rehoboth and Windy Meadows, however, and many small shelters, with chickens going in and out of them, dot the landscape instead.

Though made of steel, all the shelters are portable so they can be rotated with the chickens, and they come in three types. One is a hoop house: floorless oblong semicylinders 14 feet wide, 7–8 feet high, and 20–24 feet long, with screen panels and doors on both ends. Hoop houses are home base for meat chickens, or broilers, when they are moved from the brooder, a lighted, temperature-controlled section of a barn, onto pasture, usually at three weeks old. For the next seven weeks, at which point a flock is harvested, hoop houses provide the main protection from weather and predators. Additional protection against predators comes from the flexible, electric fencing of the paddock, and from Great Pyrenees dogs. Only sometimes at night are the broilers confined. A second type of structure has no confinement features at all. It is simply a 10- by 12-foot steel sheet with four, 2-foot-long legs attached to PVC runners. Called sleds, these supplement the hoop houses in providing the broilers with extra cover. The third type of poultry shelter is a one-room house on wheels for hens, which, like broilers, forage in paddocks on a rotating basis. Measuring 8 feet wide by 20–24 feet long and 6–7

Mike Hale and a worker mix compost outside the Windy Meadows Farm chicken-processing facility.

feet high, the hen houses have laying boxes on one side and roosting boxes on the other. They also have wire bottoms, both for aeration and to let the hens' manure fall through to the ground.

The hens are colorful breeds, reddish-brown Araucanas and Barred Rocks with black-and-white feathers and bright red heads. The broilers, like most produced in this country, are Cornish Cross, a naturally double-breasted bird with white feathers. And because chickens are omnivores with nutritional needs that foraging alone cannot consistently meet, mixtures of organic seeds and grains are part of their diet at Rehoboth and Windy Meadows, but only as a supplement to, and never a substitute for, foraging. Every day, throughout the day, the chickens are on pasture, and, with so much hunting and scratching and pecking around so many small shelters, the poultry paddocks look like thriving little villages.

The chickens' prominence is not merely visual but also economic. Broilers are the Hutchinses' main meat product, surpassed in profitability only by their Grade A raw goat milk. They raise just under ten thousand broilers a year and process them in their on-farm

Connie Hale in the kitchen with sprouting grain and rising dough.

facility. They sell most of the broilers and their other meat products and eggs year-round at the Dallas Farmers' Market, under the auspices of a marketing consortium called Texas Meats Supernatural, which Robert established in 2001 and, along with the Hales, includes one other Dallas-area meat producer. Every Friday and Saturday, members of the consortium take turns staffing the booth. Since 2005, the Coppell Farmers' Market, held Saturdays from May into November, has provided another venue for the Hutchins family, and since 2006 so has a similarly seasonal market in McKinney. The Hutchinses also operate a farm store, primarily because state law requires raw milk to be sold where it is produced, prohibiting its sale at regular retail establishments, including farmers' markets. So on Saturdays they open the store for goat milk sales and sell eggs and meat products at the same time. Selling directly to the public rather than through wholesalers, they communicate with their customers about their chemical-free, organically sustainable methods and see no advantage in organic certification.

For the Hales, broilers are the most important product, bar none, providing the family up to half its annual income. "They're our centerpiece," Mike says. The Hales raise more than ten thousand broilers a year and, like the Hutchinses, process them in their own on-farm plant. State law requires that a volume of more than ten thousand chickens be processed in a state- or USDA-licensed facility built according to certain specifications and that a health inspector attend each processing, so Mike, whom one of the Hutchins children calls a mechanical genius, designed and built their plant to comply with Texas regulations. Though the Hales have been selling their broilers and other animal products to the general public through Texas Meats Supernatural since 2002, their first major broiler customers were chefs at several white-tablecloth restaurants in Dallas, including the

French Room, the Green Room, Nana, and York Street. Mike began developing this clientele in the mid- to late nineties as part of the gradual move into full-time farming, and the chefs continue to be substantial customers. Mike makes a round of deliveries at least once a week. For the same reasons as the Hutchins family, the Hales are not certified organic.

Besides broilers, turkeys are the only animals the Hutchinses and Hales process themselves. They have beef, lamb, and pork processed at a local USDA-inspected facility, to comply with state health regulations.

Both families were drawn to pasture-based meat, dairy, and egg production largely for nutritional reasons. Though the nutritional value of organic fruit and vegetables compared with conventional is still debated, the nutritional superiority of pasture-based animal products to grain-fed is well established. Meat, milk, and eggs

Packaging chicken scraps in the processing facility of Windy Meadows Farm.

Sarah Hale leaving a Windy Meadows hoop house.

from animals raised on pasture have less total fat, less saturated fat, and fewer calories and are rich in beta-carotene, folic acid, and vitamin E. They also contain a balanced ratio of omega-3 fatty acids and omega-6's, a combination that helps prevent heart disease and strengthen the immune system. Another health benefit is lower levels of antibiotics in grass-based animal products. Pastured animals get sick far less often than grain-fed animals and are given antibiotics only when necessary to treat illness, not as a preventative routine.

"Nutrition is the driving force in what we do, after the principle of being with the children," Mike notes. And their philosophy is not to produce food that's simply better than conventional. "What we're pursuing is the best of the best," Robert elaborates. "We want the ultimate in nutrition." To achieve that, the health of the soil comes first, Mike emphasizes. "It's the key to everything. Healthy soil means good forage, healthy animals, and the most nutritious food."

The beauty of their multispecies grazing system is that it allows the Hutchins and Hale families to produce the variety of nutritious products they want and nourish the soil at the same time. They stock breeds that do well on pasture in a hot climate—Red Devon and Angus cows, for example, Nubian goats, Katahdin sheep, Berkshire pigs, Bourbon Red turkeys—and all species, through frequent, carefully controlled grazing rotations, play complementary roles in the soil-building process. Chickens and turkeys, for example, cannot forage in the tallest grasses, so cows or sheep graze an area ahead of them, cutting the grass down to size and depositing manure, which fertilizes the soil and attracts insects that the chickens eat when they are moved in. Pigs, for another example, eat an area down much as cows and sheep do but they also uproot; when appropriately directed, their uprooting activity tills areas that need it and helps control or eliminate undesirable woody plants which, left unchecked, would eventually overtake the grasses. And at Rehoboth, the goats, which naturally are browsers and prefer woody plants to grass, also help control brush and promote the growth of grasses.

Knowing when not to let animals graze a particular area is also necessary in building healthy soil and forage, and so resting paddocks are an indispensable feature of the rotation process. Like the duration of grazing periods, the length of rest depends on how rain, or the lack of it, and temperature and other weather conditions are affecting the soil, forage, and animals at any given time. "When you manage your pastures using these intensive rotational grazing methods, you keep most of your pasture resting all the time and not getting grazing pressure," Robert explains. "And so it's a great conservation of the moisture in the soil and subsoil."

Continuous composting is another essential part of soil health. Utility maintenance crews dump mounds of wood chips at both Rehoboth and Windy Meadows, and the families apply a lot of this material directly to recently grazed paddocks. Layered on top of fresh manure, it forms a sheet of compost that decomposes on site, enriching the soil without further labor. They also put a lot of the wood chips in big chainlink-paneled bins and mix them with chicken viscera and other parts. "We compost all the guts, feathers, and blood from all the chickens we kill each year," Robert says, "and you might think this would be a tremendous volume but it's not all that tremendous once it's composted. We do everything we can to keep our pastures very high in organic matter, and we've improved the grass production problem by two orders of magnitude compared to ten years ago. But still, I wish I had ten times as much compost to use. You can't have too much compost." Mike, even with a significantly smaller farm, forty acres compared with Rehoboth's three hundred, can't get enough compost either. In fact, as we arrived at Windy Meadows, our first sight of Mike found him forking chicken viscera into bins of chips and stirring the stuff.

Yet, for all the care they give the soil, as Christians they do not revere it or any other part of the natural environment, and they consider it important to differentiate themselves from people—farmers and consumers alike—who seem to worship nature. "We aren't pantheists. We believe that God created things in a certain way and put things in a certain natural order and we try to follow the purpose of creation in the order that he

Mike Hale and friend.

created it," Robert explains. "For example, ruminant animals were created with multiple stomachs to process grass and other forage. They weren't created that way to eat grain or chicken manure or feathers or the other things cattle are fed in feedlots. And so our purpose in pasturing our animals and not using antibiotics and growth hormones and not spraying toxic chemicals isn't because we don't want to harm mother nature. It's because we think God got it right when he designed the soil and the soil structure and nutrients and the way the different plants interact. So we encourage biodiversity, but everything we do and the reasons for doing it are based on our Christian worldview."

Robert and Mike think that organic agriculture has been dominated, to its detriment, by environmental extremists without a Christian perspective. "And a lot of those people see no place in the movement for Christians, but they're wrong," Robert says. "There is a growing number of Christians involved with eating grass-based meat and dairy and organic fruit and vegetables because they believe that's the most nutritious diet. I mean, the Bible says the body is the temple, and so all of us Christians want to preserve the temple in a way that's pleasing to God. Everything we do is from that perspective."

Both Robert and Mike acknowledge Joel Salatin of Polyface Farm in Swoope, Virginia, as the main inspiration and model for their farming. Salatin, who describes himself as a "full-time Christian libertarian capitalist environmentalist farmer," has a journalism degree from Bob Jones University and not only raises meat animals on pasture but writes extensively about how others can do the same. He and his wife and two children got started in the early eighties, and their methods proved to be increasingly profitable. His numerous books and a regular column in the monthly magazine *Stockman Grass Farmer* draw on his family's successful experience and have made him a nationally influential figure. Polyface Farm comprises 550 total acres, 450 of which is wooded and the rest in pasture. The contrasting sizes and other features of Rehoboth and Windy Meadows suggest how widely adaptable Salatin's model is. "We're grass farmers," Mike says, which, like Salatin, is what many farmers who reject conventional grain-based meat production call themselves, Christian or not. "We're grass farmers going beyond organic," he adds, and notes

Ruth and Abigail Hutchins with chickens and goats, Rehoboth Ranch.

that the distinction is necessary because many "merely organic" producers feed their animals grain and may or may not actually put them on pasture. Robert likes another name as well. "My family and I tell people we're in the protein business," he says, "because we raise all these different species of animals for meat, plus goats for milk, and hens for eggs."

Robert Hutchins is a big, tall man with an "I'm in charge here" manner softened by old-fashioned politeness. He was born in Greenville in 1953, into a family with a history in the area dating from the 1800s. "But nobody was ever full time in agriculture until me," he says. His mother was a homemaker, and his father worked in the defense industry, at the E Systems and Raytheon plants where Robert himself worked from 1980 until 2000. Robert's father also served in the Texas House of Representatives, and so did his grandfather. Robert is the second of five children and the first son. His family lived on a farm but didn't farm. "We had a few cows and a vegetable garden," he says, "but that's kind of what you did if you lived in the country."

Nancy was also born in 1953, in Tripoli, Libya, where her father, a master sergeant in the air force, was stationed. In 1967, he retired and settled in Greenville with his wife and Nancy and her two younger sisters. The family happened to join the Methodist church the Hutchins family attended, and Nancy and Robert met in Sunday school. They were fourteen. Eight years later, in 1975, they graduated from college, married, and moved to the San Diego area. Robert, just out of the Naval Academy, was assigned to work in surface warfare systems, and they lived in Southern California for five years. As they set up their first household and domestic routines, one of the things they enjoyed most was buying and preparing fresh, locally produced food and getting lots of exercise. "All of our friends and neighbors were really into food and physical activity, and we got the bug. It's just part of the culture on the West Coast, farmers' markets with all kinds of organic food, raw milk in grocery stores, and people cycling, surfing, running," Robert recalls a little wistfully. "I was running thirty miles a week, and we were eating all this great food every day."

Robert and Nancy took this culture for granted until they returned to Greenville to live, in 1980. They had two children by then and wanted them to grow up near both sets of grandparents. "It hit us like a brick that the way we ate in California just isn't part of the culture here. Growing up in a place, you can't see what you're missing, but when you move out and then come back, it hits you. The organic fruits and vegetables and meats and dairy products we were used to in California, our whole way of eating, really, was beyond the lunatic fringe here," Robert recalls. "I mean, it wasn't even on the radar scope." They bought twenty-one acres adjacent to the farm where Robert had grown up and his parents still lived and built a house. To eat even slightly as well as they did in California, they started growing their own organic vegetables and keeping laying hens.

The hens were Nancy's idea. Her grandmother lived in a little Nebraska town and kept chickens in her backyard, and Nancy and her sisters had fun gathering eggs when they visited. "So when Robert started talking about moving back here and living in the country, I said, 'Well, if I can have chickens that lay eggs,'" Nancy remembers, laughing. "Little did I know where that was going to lead!" She's a large, poised woman with warm eyes and a warm voice. Despite the pleasures of Southern California, she didn't shy away from returning to Greenville or from the many big changes that followed. "I'm a go-with-the-flow, take-it-as-it-comes kind of person," she explains.

Robert and Nancy Hutchins.

Living in the Greenville countryside, they soon found themselves wanting to produce more of their own food than just eggs and vegetables, so they expanded from hens to a few flocks of broilers, a few cattle, and an occasional couple of lambs. "And because of the interests we brought back from the West Coast, we started taking our animals and pastures off of chemicals," Robert says. "And at that time we thought well, okay, chemical residue is a bad thing. If you can just get rid of chemical residue, then you'll have healthy meat." But their thinking soon moved beyond that. Robert began reading about organic agriculture, and this reading, along with the decision to homeschool their children, only intensified their interest in alternative ways of doing things.

Robert realized that to learn how to produce the kind of food he and Nancy wanted for their family he would have to teach himself. "I didn't have any formal education in agriculture, and I knew from what I'd read through conventional channels that I didn't want any," he says. "I mean, Texas A&M–Commerce is only fifteen miles away, but I didn't want to learn what they were teaching. I wanted to learn things they weren't teaching

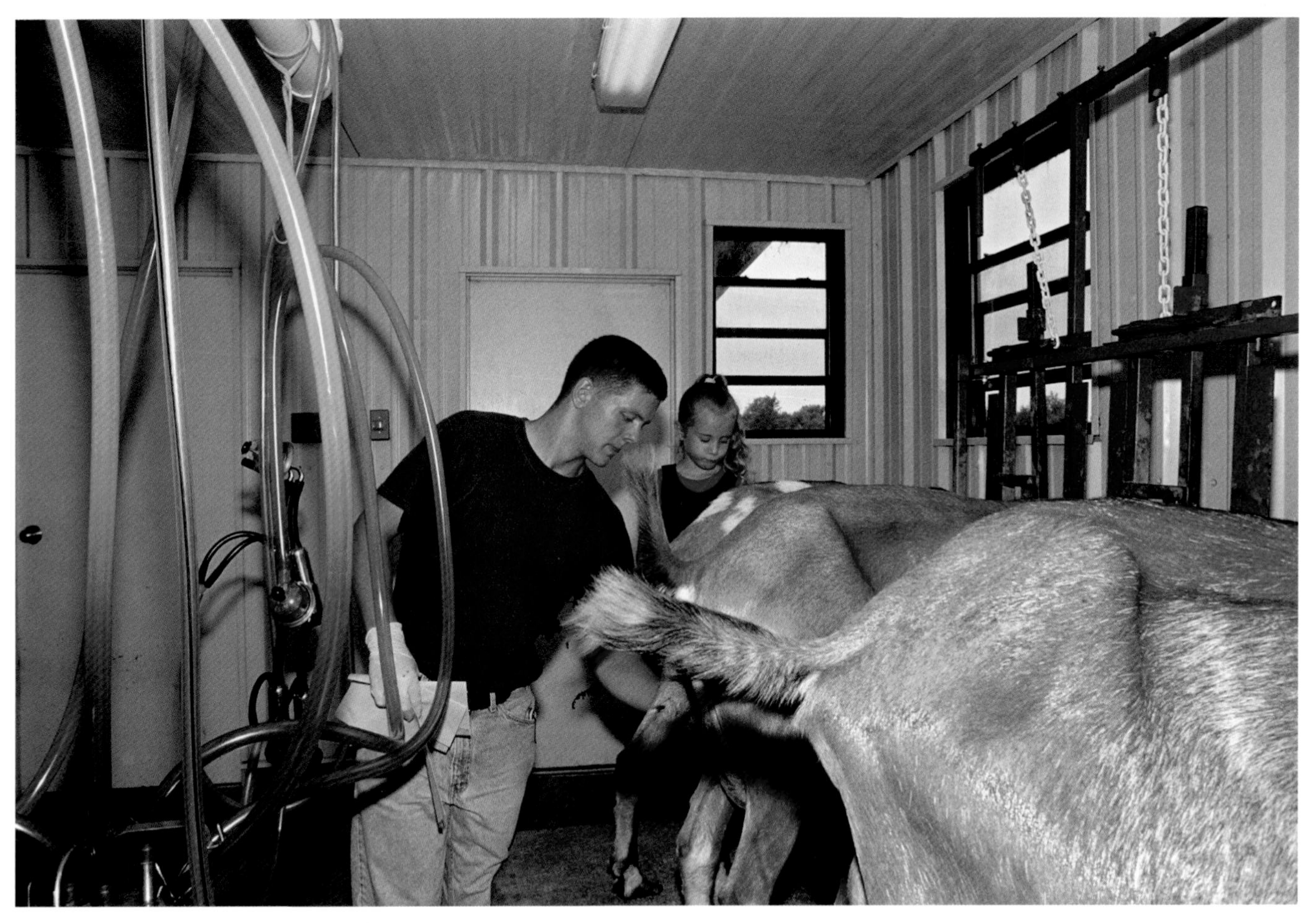

Mark and Ruth Hutchins prepare for milking.

over there. So it had to be self-education. I started going to alternative ag seminars, subscribing to *Stockman Grass Farmer* and other alternative ag periodicals, and reading Salatin's books and every alternative ag book I could get my hands on. And we began experimenting with different methods, raising chickens in a free range and using intensive rotational grazing with the cattle and just learning along."

By 1988, eight years after returning to Greenville, the Hutchins family decided to go a little beyond producing just for themselves and started selling a small volume of meat products locally. Robert and Nancy had determined by then that they wanted farming to be their full-time livelihood, with Robert resigning from his job and the whole family working together. Robert even informed his boss that this was his goal, as soon as it seemed economically feasible. From their eight years of experience, he and Nancy felt that they had not only the desire but the ability to make this change.

In addition to schooling their children at home, they had begun holding church at home, too. Institutional Christianity had come to seem shallow and irrelevant, because it undermined rather than highlighted the primacy of the Bible as a guide to daily life. By studying scripture and worshiping at home, they could restore the Bible to its proper role. "Everything we do is based on what we see to be the truth from the scriptures," Robert says, "and I had begun to understand directly from the pages of scripture the importance of my role as father and husband and the priority that I'm supposed to give to them, and I thought God wanted me at home working with my family. I think God wants us dependent on him, not on our paycheck."

But deciding to do without a paycheck took twelve more years, until 2000. "I had a six-figure salary and didn't want to give it up," Robert says. "It was hard to pry my fingers off that. And even now, when people ask me if I miss my old job," he chuckles, "I say only every other Friday." He might have brought himself to resign sooner, he reflects, had it not been for a couple of factors involving his extended family. One was the weight of his father's advice. "I told him I wanted to work with my family and do things I enjoy right now," Robert remembers, "to live life every day and not just expect to live it sometime in the future. And he said, 'Well, just go slow and keep it small scale and work your way into it. Don't quit your job right now, but start producing, start selling, and see if it grows.' And that's the approach we took."

A second factor was the death of Robert's father and mother in the early nineties and subsequent disagreements among Robert and his siblings regarding the family farm that adjoined Robert and Nancy's twenty-one acres. "We thought the family farm would be ours and that we'd be staying there, but it wasn't left to just us," Robert says. "And then there was also conflict about farming. Everybody but us wanted to spray weeds with herbicides, feed animals grain and give them antibiotics, the whole conventional deal. We realized it would be best for us to sell our house and acreage and find another place."

So in 1995 they purchased what would become Rehoboth Ranch and moved there. As they were considering what to name their place, they happened to read the Genesis account of Isaac, living among Philistines, reviving a well that had once been his father Abraham's. Isaac named the well Rehoboth and said, "For now the Lord hath made room for us, and we shall be fruitful in the land." They found the name appropriate and decided to couple it not with "farm" but "ranch." Robert seems amused, telling the story. "Everybody raised in East Texas knows there's no such thing as a ranch here. A ranch is a West Texas thing," he says. "Here you have farms. You raise crops, you have a farm. You raise animals, you have a farm. You raise strawberries, you have a farm. I mean, it doesn't matter. In East Texas it's a farm. But our customers are city people, and when city people buy meat they want it to be from a ranch."

Rehoboth encompasses about 225 acres of pasture, including both native and nonnative grasses, and about seventy-five acres of woods, mainly pecan, red oak and other oaks, hickory, and ash. The woods provide browse for goats and also habitat for such animals as purple martins, barn swallows, and bats. The woods also provide the Hutchins family with fuel for the heating stove in their house. The northern boundary is shaped by the headwaters of the Sabine. "When there actually is water!" Robert laughs. "It doesn't run year round here. It's a little drainage most of the time." The soil is mostly sandy loam.

With Robert's income, they were able to purchase the land outright and to eliminate all their debts by 2000, when Robert resigned from his job. "We felt very blessed to be able to do that," Robert says. "Plus, being with the company for twenty years, I got twenty weeks of severance pay, and that helped." They anticipated a profit after two years, but it took three. "Everything had to go up a notch in volume, and when you increase production twofold, threefold, fourfold, fivefold, there's a learning curve and there's expenses," Robert explains. Six years into full-time farming, their profit margin is adequate but smaller than they'd like. "We keep our bills paid," he says, "but we're not to the comfortable level yet, though our type of operation can eventually provide a comfortable lifestyle."

Adjusting to a dramatic reduction in income has been eased not only by the absence of debt but also because, even when Robert earned a high salary, the Hutchins family did not spend it the way many affluent Americans do. "We used a lot of money to buy our first farm and then this one," Robert says, "instead of big, fancy houses." And Nancy notes, "We always drove older cars even when we could have bought new ones." They didn't have typical health insurance, either, and still don't, but they participate in a national Christian cost-sharing organization, as they have for years. They find it to be more economical and ethical than other options. "It's not insurance but an organization that follows biblical principles of Christians caring for each other," Robert explains. "Each month people who have health bills submit them to the organization and then the organization notifies the members who you should send your check to. So we'll get a letter telling us something like this month we need to send a check for $195 to the Smith family in Jonesboro, Arkansas, because Mr. Smith has a broken leg, or whatever."

Getting along with less money has been easier for the family than dealing with the workloads necessary for economic viability. They hire no outside help, and Robert and Nancy and the ten children still living at home all share in doing whatever needs to be done. New jobs arise and ongoing ones change as ages and abilities change. Building a Grade A goat dairy and getting licensed by the state in 2003 was a three-year project, timed for completion when two children would be old enough, in their teens, to run it. Twenty goats are mechanically milked four at a time twice a day. Processing the raw milk and managing the goats' foraging is the single most labor-intensive job at Rehoboth. Taking care of the broilers is the second most demanding, and a fifteen-year-old son takes care of them, though on chicken-processing days everyone old enough to help is involved.

As children's abilities and jobs change, some of Nancy and Robert's jobs do, too. Until recently Nancy conducted the homeschooling, managed the household, prepared the meals, and made much of the family's clothing, but two older daughters now do most of the teaching, cooking, and sewing. Though Nancy continues to oversee teaching and housekeeping, her newer duties include running most of the errands and helping sell at farmers' markets. Robert manages the production and marketing. He once did the bookkeeping and handled e-mail communications with customers, but the oldest daughter now does most of this and also keeps track of the inventory in the walk-in freezers. A sixteen-year-old daughter tends the vegetable garden, the younger children gather eggs, and so on. No one gets paid. "Everyone that's old enough has their own money-making opportunities," Robert explains. One daughter sews for several people, and another one cleans a couple of houses.

The seasonality of much of the production helps lighten the workloads, especially broiler production, which is suspended from about Thanksgiving until March. The family also suspends dairy operation for about one month of each year, to give the children responsible for it a break. Regular recreation also helps. During warm weather, they swim once a week in their large pond at Rehoboth. But their favorite recreational

Red Devon and Angus cattle in the Rehoboth Ranch shade.

activity is whitewater canoeing and kayaking, which they do as often as they can, usually in Oklahoma in one of the tributaries of Broken Bow Reservoir. "We never go for a long period of time," Robert explains. "Usually just a day. We'll get up at three A.M. and milk the goats, be gone by five A.M., on the river by ten and off by five, and home by midnight."

Some of the older children note that it's not so much the volume of work that sometimes bothers them as the lack of variety in responsibilities and routines. Free time allows a change in pace and opportunities for developing individual interests, yet they typically choose an activity related to their work. Mark, a lean, adult-like seventeen-year-old, and one of the two children responsible for the dairy, points out the irony. "I'll be wishing for more free time, but what do I do? Work! I like researching things," he says, "so in my spare time I often write little papers on some current interest. It might be a an agricultural issue, or a political one, or one that's both, whatever I'm thinking a lot about at the time." Stephen, who is thirteen and has a ready smile, takes digital photographs. His subjects are the grasses and flowers, birds and butterflies, and snakes and reptiles that he has learned to pay attention to just in the course of working with the various farm animals. Elizabeth, twenty-one, does dairy work with Mark and also helps cook meals and sew clothes for the family. When she has leisure time, she likes going to her favorite fabric store in Dallas, especially if there's a sale, and stocking up for whatever she will make next. "I like finishing a good garment," she says, smiling as if amused that her free time, like Stephen and Mark's, somehow keeps her close to her usual work.

"I think most of us have learned to keep things in perspective," Elizabeth goes on to reflect. She has dark brown eyes and hair and her mother's poised warmth. "Everything's not perfect, and things get frustrating some days, but, you know, everyone has frustrating things in their lives. And when you start complaining about how much work you have to do, well, if you look around at other families and other people's lifestyles, then you realize how blessed you really are. And our customers' and friends' appreciation really helps keep things in perspective, too. So often someone comes up to us at a market or comes out here and says, 'Oh how wonderful! I would like to be able to do the things you do!'"

Still, whether any of the Hutchins children will choose to farm for a living is an open question. The two oldest sons, who no longer live at Rehoboth, so far are pursuing other work. Matthew, the Hutchinses' firstborn, studied civil engineering in college for a couple of years but decided that a sedentary life at the computer did not appeal to him, so he developed his own construction business. He works mainly in Dallas, but he and his wife and two young children live in rural Greenville and see the larger Hutchins family often. "He was always our mechanical person," Robert explains, "and he's still on call, consults with us. Truck breaks, I call him and say here's the symptoms, what do you think, and he tells me what to do." Matthew and his family also participate in the home church that meets at Rehoboth, and most of Matthew's ten full-time employees are young men associated with that Christian community. Brian, the second-oldest son, received enough college credits from tests that he skipped college altogether and went straight into studying law, through a distance-learning program of Oak Brook School of Law, based in Chicago. After completing his courses, he worked as an intern for an attorney in Minneapolis, and at the time of our visit he was about to move to California because Oak Brook graduates take the California bar exam. "After that," Robert says, "who knows?"

Among the oldest of the ten children still at home, Mark and Elizabeth speculate about the future along similar lines. "We all want to stay close by and live in the area, for sure," Mark says. "But no one knows for sure whether or not they want to stay on the farm, running it. We'll just see how God works that out. But if I bought a house today, it would be down the road. I want to continue living in the country and raising my own animals, whether I farm for a living or not." Elizabeth expresses the same desire and adds, "I hope I never have to live in town. I think I would be bored to tears."

"I think they see the value of producing their own food versus what's otherwise available out there," Nancy observes, and Mark affirms this. "We learn so much ourselves and try to teach our customers," he says. "We tell them why it matters to eat food that's produced the way we do it, and we say, 'If you can, raise your own chickens, your own eggs, and get a cow,' that type of thing."

As the children grow up and the Rehoboth household grows smaller, it will require less income and Robert and Nancy will reduce the scale of production. "You don't have children for your work," Nancy explains, "you have work for your children." Robert points out, however, that their food preferences and farming methods require a certain level of diversity, so they do not envision raising just one or two species. Nor do they envision ever retiring from family farming. They hope that at least one of the children eventually chooses to take over the business and keep it going, with Robert and Nancy helping. "All we need is just one out of the twelve!" Robert laughs.

Whatever the future holds, the Hutchins family thinks less about it than about the challenges and pleasures of the present. "The most surprising thing to me is how hard doing all this is," Robert admits. "All the years we spent producing small scale made me think that once I had full time to devote to work on the farm, then after about a year I'd have my to-do list all knocked out. I'd be sitting on the front porch every day about ten o'clock drinking iced tea, you know, and have all this time to spend with my family, time we wouldn't be working. And that just hasn't been the case. For every one thing I mark off my to-do list, three or four things get added to it and I never get caught up."

He confesses that he also has had to learn to be more patient. This part of our conversation takes place during lunch with Robert and Nancy and eight of the children. We're seated around a rectangular wooden table in the spacious main room of the house—its heart, the dining area, at the center, an open kitchen on one side, an open living area furnished with rocking chairs and pews on the other. Elizabeth has grilled chicken and squash and eggplant, platters of it, and made pitchers of iced tea. Ruth and Abigail, blonde, seven-year-old twins, keep our glasses filled. All the children nod attentively and smile as Robert speaks. "I was used to dealing with adults, not children, and when I said something, when I wanted something done, it got done. I mean, it was just unthinkable, in my jobs before, to say we need to do thus and such and then find out it wasn't done," Robert says. "So I had to get used to, well, the children are not adults, there's still training to go on and just because I tell someone to do something doesn't necessarily mean it'll happen. That was a big adjustment." He pauses and smiles along with everyone else. "Not that it's a done deal," he adds. "There are numerous opportunities still to learn."

"For all of us," Elizabeth puts in, and once again everyone smiles knowingly at each other.

"So in every way," Robert continues, "farming is as hard as any job I've ever had, and the economic rewards are not nearly as great as if I were still applying my time and talents in the business world, but the nontangible rewards are immeasurable. Our underlying purpose for what we do is to be able to raise our children with the right kind of character, with training rooted in the Bible. In our community, we say we want our children to have the character of Christ, and we think this is good

Bourbon Red turkeys foraging at Rehoboth Ranch.

training ground." But he interjects that agriculture is not the only or necessarily even the best training ground. "Some very fundamentalist Christians think that the scriptures teach that agriculture is the only valid work, but I don't subscribe to that at all," he says. "We know a family with a window-washing business. Working together as a family is what matters. Doesn't have to be agriculture."

Robert acknowledges, however, that agriculture does present opportunities for ministering to the public that other types of family businesses may not, and that a growing sense of ministry motivates his family. "The longer we farm, the more we see it as a form of ministry, because we're providing people with nutritionally superior food that is not easily available from other sources," he explains. "We make a connection between the rise in lifestyle diseases and nutrition and we want to help other people make that connection and to provide them with food that helps prevent disease and improve health."

Reaching out to more people and keeping them as customers depends on differentiating the quality of their products and their methods from both conventional agriculture and industrial organic. "Customers who take the time to drive to the farm for milk, meat, and eggs already understand the difference and are committed in their support, but customers at farmers' markets are a different story," Robert finds. They may or may not know much about nutrition or agriculture, and turning them into loyal customers means not only offering excellent products but also engaging them in conversation and giving them printed information about small-scale, pasture-based farming. "When I'm feeling especially tired," Robert laughs, "I say I feel like we're educating Dallas one consumer at a time. In this business, you just sign up for it!"

With Wal-Mart getting into organics, Robert believes public education will be even harder and more important than ever. He thinks Wal-Mart's marketing power will spur industrial organic production throughout the world, much of it by multinational corporations. And these corporate giants, he expects, will influence the USDA to lower its organic standards even further than loopholes like the "access to pasture" rule already do. "If big companies could do organic farming right, people like us would be out of business. But I have no confidence they'll ever do it right," Robert says. "So as they continue to corrupt the organic standards, we'll have to refine how we differentiate our products. I don't think we can survive if the only differentiation is we're local and we're a family and we do things organically. Now we're going to have to throw in "sustainably" and talk a lot more about raising food organically and sustainably. Doing things sustainably equals good stewardship of the earth's resources, and Wal-Mart and its industrial organic suppliers aren't doing that. Farmers like us are, and we've got to get people to understand and care about that and support this kind of farming."

Robert would like consumers to stop thinking of themselves as consumers and instead consider themselves co-producers. This is the term advocated by Carlo Petrini, founder of Slow Food, an international organization that promotes sustainable farming and local food systems and communities. "Co-producers" resonated with Robert as soon as he heard it, in October 2004, when he and Nancy visited Turin, Italy, to participate in Slow Food's first gathering of small food producers from around the world. "Farmers alone can't advance organic and sustainable agriculture," Robert says. "Consumers have to drive it, too. And the more people we can get to think not just about what they're putting in their mouths but about the kind of agriculture that they are funding with every dollar they spend—to think of themselves as co-producers—then the likelier they'll be to help make it possible."

And he doesn't mean just by purchasing products. Customers as co-producers can also help by actively working to change policy and regulations that make it difficult, sometimes impossible, for small farmers to make a decent living—laws, for example, that limit how many chickens a farmer can process on a farm without building an expensive processing facility, or laws that prohibit the sale of raw milk at retail outlets, limiting farmers' venues and forcing customers to make special drives to farms producing it. "If people think of themselves as co-producers," Robert says, "they'd learn to see that whatever affects the small, local farmer they're buying from affects them, too, affects what foods they'll be able to eat and what they have to do to get it."

The Hutchins family spends thousands of dollars a year and a great deal of time to comply with regulations for disposing of wastewater, processing poultry, and producing Grade A raw goat milk. The dairy regulations are the most onerous. They derive from conditions at industrial cow dairies producing vast quantities of pasteurized milk and overlook not only the very different economic scale of small dairies but also the very different ecologies of the facilities, including the biochemistry of goats. At industrial dairies, hundreds of confined, grain-fed cows are milked at a time, standing in long assembly lines of stalls. Few people are on the scene, and those who are cannot exercise anything like the vigilance the Huchinses do, with two people milking only four goats at a time—keeping all the manure washed and drained from each stall as it drops, for example, to prevent accumulation and spattering onto the animals' udders, and, as a further guard against contamination, manually disinfecting each udder immediately before attaching the suction apparatus. Another sharp contrast is that testing for diseased animals is not required for dairy herds whose milk is pasteurized, whereas goats and cows at licensed raw-milk dairies must be tested regularly. "We just don't have the same sanitation issues as big dairies," Robert says. "We're small and clean, and animals raised on pasture aren't prone to the same diseases as those confined. Plus, goats aren't cows. They can have certain sorts of cell counts that are twice the regulatory limit for cows yet they're perfectly healthy. But the regulations don't allow for biological differences like that. They aren't based on science. They're based on problems with large-scale dairies that don't have any real bearing on dairies like ours, but that's how the regulations go. And our license to milk a small number of goats costs twice as much as a license to milk a large number of cows."

In limiting the sale of raw milk to the farm where it is produced, the regulations also reinforce the conventional assumption that raw milk is by definition tainted with harmful bacteria and therefore unsafe. Federal law, for example, prohibits the sale of raw milk across state lines, and the only states that permit its sale in retail stores are Arizona, California, Connecticut, Maine, New Mexico, Pennsylvania, and South Carolina. Despite the restrictions, however, many people prefer raw milk to pasteurized for its flavor or because they think it is more nutritious, or both. New York University nutritionist and author Marion Nestle writes that there is no evidence from food composition research that pasteurization diminishes nutritional value, but others, especially in alternative food communities, disagree. These include Sally Fallon, head of the Weston A. Price Foundation, a nonprofit organization devoted to alternative nutrition research and education, and Nina Planck, food journalist, founder and director of the London Farmers' Markets and former director of the New York City Greenmarket. Certainly the Hutchins family is among the dissenters. "Pasteurized milk is dead milk," Robert says, summing up the shared attitude. "Pasteurization kills enzymes and other beneficial organisms along with harmful things."

Detractors of pasteurization generally do not dispute that it is necessary when milk from thousands of cows is combined and then packaged and distributed throughout the nation, but they do want to dispel notions that this production system is best, that pasteurization provides fail-proof protection from pathogens, and that raw milk is always risky. They point out that pasteurized milk is continuously subject to contamination during handling and packaging, transport and storage, and in cheese making, and they cite the many reports of *Listeria*, *Salmonella*, other pathogens, and even chemical toxins turning up in pasteurized milk and cheese. They argue that raw milk produced according to Grade A standards is far less risky than industrially produced pasteurized milk, as well as being tastier and more nutritious, and they strenuously object to laws that prevent its sale in grocery stores and other mainstream retail venues.

Robert finds this marketing restriction the single most onerous of all the dairy regulations. "Our milk is so much cleaner than pasteurized because the sanitation standards are far higher for us than for conventional dairies. I mean, you can find more bacteria in a jug of pasteurized at the grocery store than you can in a jug of raw milk that we sell here," he says. "Sell here and by Texas law can't legally sell anywhere else!" he emphasizes, his tone more than a little exasperated. "Even though we have a dairy license and the state does the testing and knows the milk is safe! You can take pasteurized milk somewhere but you can't take raw milk anywhere! Now the only purpose of that is to discourage people from drinking raw milk."

And to discourage family farmers from producing it, for, as Wendell Berry has recurrently noted, dairy regulations have been at least as effective in eliminating family dairies as they have been in reducing milk-borne pathogens. Robert thinks that to relieve small farmers and their customers, or co-producers, of these and similar hindrances there ought to be what he dubs a "farmer-with-a-freezer" law—"a law," he elaborates, "that says if you're a small-scale farmer, which I would define as being under a million dollars a year in gross revenue, then you're exempt from all regulatory intrusions from all government agencies at all levels. And all you have to do is put a sign at your front gate and product label that says, 'This product was produced by a small-scale farm, and has not been subject to any regulation inspection, oversight, or government involvement whatsoever. Buy at your own risk.'"

"I would be happy to do that. Happy to do that!" Robert adds. "And if I ever get some spare time," he muses, "I just might do something about getting that law."

It's a plausible threat.

In 2000, the state health department notified Robert that, to eliminate an ambiguity in existing regulations, it would soon prohibit any on-farm poultry processing. This change would have forced farmers to transport their poultry to off-farm processing facilities, incurring costs that would put most small commercial producers out of business. Robert and his law student son contacted several state representatives and senators to get an amendment attached to an existing bill to eliminate the regulatory ambiguity and clearly permit on-farm processing. While the amendment was pending, they sent e-mails to their customers and others' customers, urging them to let their legislators know they favored the amendment. The efforts worked. The right to on-farm poultry processing was preserved.

"At one point our grassroots uprising was so strong that the state senator who had the responsibility for the bill had his staffer phone me," Robert recalls with relish. "'Tell your people to quit calling us,' he told me. 'We've heard enough, we got it, we're going to get it done. Your people can quit calling now!'"

Windy Meadows, true to its name, is mostly open fields of grass. Nonnative Dallis grass and Bermuda grass predominate, but Mike has noticed several native clovers coming back. Mottes of pecans, bois d'arcs, post oaks, honey locusts, and native persimmons dot the pastures and fringe segments of the farm's perimeter. The soil, like Rehoboth's, is sandy loam.

The farm got its name from Connie, a city girl personified according to Mike, when they first met. She was born in Ft. Monmouth, New Jersey, in 1955. Her father, like Nancy Hutchins's, had a career in the air force. Their family, which included a son older than Connie and one younger, lived in urban areas on both the East and West Coasts, then Austin, and finally Plano, where Connie graduated from high school in 1973. "Even though I grew up in cities, I was always a tomboy," Connie says. She is short and dark, with shiny salt-and-pepper hair, and is an eager yet focused talker. "I always spent as much time as I could doing things outside, especially after we moved to Austin." Somewhere along the way, she developed an interest in geology and chose it as her major when she enrolled in East Texas State University (now Texas A&M–Commerce). Living in Commerce put her in a country town for the first time and, through Mike, introduced her to some rural routines. They met in an honors chemistry class her first semester. Mike was a sophomore.

Mike was born in Commerce, in 1954, the fifth of six children, and reared in town. But both his parents grew up on farms, and his father, even after he moved to town and went to work for a railroad company, raised cattle on rented pastures. "Always, we had cattle," Mike says, "and the whole family would drive out to the country to see after the cows." Mike was still seeing after his father's cows when he and Connie started dating, and she went with him and helped him count heads, check the fences and forage, and do other chores. The farm that Mike's father was leasing at the time was the farm that Mike and Connie, without any idea of it then, fifteen years later would purchase. "Every time Connie and I would came out here, she always said how windy it was," explains Mike, an affable man with receding, reddish-brown hair. "'It's just so windy out here!' she'd say, and that came back to us as we were deciding to buy it."

Before reaching that point, however, they took a long detour to Plano, where they both got teaching jobs after marrying and graduating from college in 1977. Connie taught until their first child was born, in 1982, and then became a full-time homemaker. As their family grew, they wanted to live close enough to both sets of grandparents so that more or less equal interaction would be possible. With Connie's parents still in Plano and Mike's in Commerce, they determined that Greenville would put them about an hour away from each place, and so Mike found a job there in 1986. They rented a house in town but soon felt the pull of the country and began looking around for acreage. Two years later, in the spring of 1988, they found what they wanted, the forty acres they had been so familiar with in college.

They got loans for the land and for building materials, and Mike, skilled in carpentry and all other aspects of construction, spent the summer building a two-story house. By fall, when they and their first three children moved in, Mike's father was renting the pastures once again for his cattle, and they got a few chickens for eggs. For a long time, that was as much agricultural activity as they could imagine on the place. They had chosen to live in the country for the space and physical freedom that rural living affords, not because they intended to farm. Farming, even had they been interested, required large acreages, big facilities and machines, and the kind of money, short of borrowing, they would never have. Or so, like most people, they assumed. But in 1996, Robert Hutchins, whom they had met through homeschooling circles ten years earlier, introduced them to a different concept of farming. He gave Mike a copy of Joel Salatin's book *Pastured Poultry Profits*, and this, along with the small-scale commercial production the Hutchins family was already doing, piqued Mike's curiosity. "That book and Robert got me thinking about farming for the first time," Mike says.

Elizabeth Hutchins and younger children prepare lunch in the Rehoboth Ranch farmhouse.

Mike enjoyed teaching but, the more Salatin books he read and the more he talked with Robert, the more taken he was with the idea of an agricultural livelihood that could involve the whole family and provide them with nutritionally superior food at the same time. Connie agreed. It seemed of a piece with homeschooling and with the family's healthful eating habits. Connie and Mike had become interested in nutrition even before they had children, back in the seventies. The popular notion of homesteading—no matter that they were not actually homesteading—first cued them in to eating natural food and preparing it from scratch. "That can-do attitude appealed to me," Connie recalls, "and I started doing research about the most nutritious fruits and vegetables and grains to eat, the best eggs, the best dairy products, and I made notes and clipped and compiled articles and put everything in three-ring binders. And along with routine meal preparation, I got into things like making yogurt and kefir and grinding grains and making multigrain bread." The more conscientious she and Mike were about their diets, the better they felt and the more they saw nutrition, not conventional medical care, as the key to health. "Conventional medicine really just manages sickness," Connie says. "It doesn't make people healthy. Only eating well does that."

Discovering the nutritional benefits of pastured meat products opened up a whole new way to improve upon the careful eating they had always done. But was grass farming on merely forty acres a feasible way to support a family of eight children and two adults? And, if so, could they get started without adding to the debt—at one point more than $100,000—that they were in the painstaking process of getting out of? These debts had been necessary to provide the home they wanted for their family, but most other kinds of debts were not, they thought, no matter how routinely most Americans took them on. And so they had made it a primary goal not only to eliminate their debt but also to do their utmost not to incur any more and unnecessarily tighten a budget they figured would always be tight. Their budding desire to farm was not going to budge them from this financial goal.

"To go into farming without going into debt," Connie explains, "meant that we first had to see how we could spend less on our daily needs. We started asking ourselves *why* we were doing things the way we were and did we *need* to do them. Everything was brought under the magnifying glass." This scrutiny resulted in several changes. Nobody went to town for haircuts anymore; Connie became the haircutter. She also started making even more of the family's clothing than she had been making or, when cheaper, buying from thrift shops. And though they rarely ate out or went to movies for entertainment, they began going even less and instead invited other families over more often. Vacations, too, never regular occurrences, became even less frequent. And for medical insurance, Mike dropped the increasingly expensive coverage available through his teaching job and, like the Hutchinses, they joined a Christian medical cost-sharing organization. They also reconsidered the spiritual value of participating in institutional Christianity and their financial contributions to it. They decided that when their current responsibilities in their Presbyterian congregation ended they would become part of the home church that Robert had been instrumental in organizing a few years before.

Practicing a more stringent frugality than ever, the Hales gained enough confidence about eliminating debt that by 1998 they decided to move slowly and experimentally toward commercial grass farming. They would develop one small phase of their projected business at a time, and Mike would keep teaching until they could be reasonably certain that they could produce enough to support the family. "We started with a little core of things that the children wanted to do and were capable of doing," Mike explains. "We got more chickens for eggs, and we got some sheep, and a couple of milk cows for raw milk just for the family, not as part of our commercial farming plan." Everyone enjoyed working with the animals, and so they added some beef cows, and Mike constructed their first broiler shelters and they got their first flocks. By 2001, they felt practiced enough at multispecies rotational grazing that they began marketing their products on a small scale. Their eldest child, David, then nineteen, managed the animals and production, while Mike, still teaching, concentrated on developing the clientele, focusing first on Dallas chefs.

"What I'd do is get the list of restaurants that comes out in the Dallas paper every Friday," Mike says, "and I'd get on the phone and call up the white-tablecloth places, introduce myself, and ask if they were interested in putting pastured chickens on their menu. And if they said yes, I asked if I could bring them a couple to try for free and compare with whatever chickens they were serving, and make an appointment to deliver them. After they had time to try them, I'd go back and talk with them, and I never had a chef that tried our chickens that didn't want them." He notes, however, that some chefs' budgets prevent them from becoming clients, but by persisting with his personal approach he secured the number of restaurant accounts he had aimed for. By 2002, he could safely enough predict that this venue, together with the Texas Meats Supernatural venue Robert opened that year at the Dallas Farmers' Market, would provide enough income that in 2003 he could stop teaching. In advance of that point, during the spring of 2002, he worked with David to slowly increase the volume of all their products and spent weekends and afternoons constructing their poultry-processing facility.

Though they were completely out of debt by then and, given Mike's skills, able to build the facility without borrowing money, Connie recalls it as a period of anxiety for her. "Building the processing plant made our decision to farm full time seem suddenly permanent," she explains, "like something that, say, next week we couldn't decide not to do. And we were only just out of debt, and what if farming put us back in?" Mike had similar trepidations, especially when 2003 rolled around and he resigned. "It was scary," he admits. "Very scary!"

But three full years into full-time commercial production at the time of our visit, the Hales remain debt free. "The miracle," Connie marvels, "is that we can do what we're doing on only forty acres." All eight children live at home, though David commutes each weekday to Dallas, where he supervises the maintenance of a downtown skyscraper. In his off hours, he helps Mike with the animal rotations, and on a small, rented plot not far from home he also operates his own commercial hatchery. Windy Meadows and Rehoboth are his main customers but not the only ones. He may eventually purchase a farm of his own yet for now has no definite plans. Three much younger sons take care of the ten Great Pyrenees dogs guarding the pastures and tend the hens and gather eggs. Four teenage daughters divide among themselves a variety of tasks: tending the milk cows and operating the milking machine, monitoring the sheep, helping Connie in the kitchen, and helping Mike at the Dallas Farmers' Market. On processing days, as at Rehoboth, everyone old enough pitches in. Mike is in charge of production, marketing, and all the paperwork except for poultry processing, which Connie handles along with managing household chores.

Connie also has a couple of part-time "cottage industries" going. She conducts nutrition seminars in their home, based largely on the theories and advice of the Weston A. Price Foundation, which promotes the health benefits of eating "nutrient-dense" natural foods including pasture-based meats, raw milk, whole grains, and lacto-fermented vegetables (a traditional preserving process involving whey). Connie discovered the organization online a few years ago while preparing nutrition lessons for her children. The organization's emphasis on preventing disease nutritionally rather than treating it medically complemented and expanded her and her family's own dietary thinking and habits so compellingly that she joined a local chapter and soon became a leader and teacher. With the help of an older daughter, Connie also makes custom wedding cakes. The activities supplement the family's farming income, but in a small way. Connie does these things because she enjoys socializing and, in the seminars, sharing her ongoing learning about diet and health—and also because she sees it as she and Mike also see their farming as a ministry. "Because we kept to our course and were faithful to not listening to other voices but were faithful to the course we felt God was leading us through," she says, "now we have customers and so many other people wanting to learn from us. And we can share, and I'm so thankful."

Mike considers farming the most challenging work he has ever done. "We're really running not one business here but three. One is producing, and that in itself is several different productions, and then there's processing, and then marketing," he points out. "You don't realize how complex it is until you get in the middle of it and try to make everything work. Outside of trusting the Lord, I'm not sure I could take this all on." Not surprising for a former math teacher, Mike thinks of farming as a mathematical equation. "I'm always factoring in all the parts, always trying to figure out how the failure of one part would affect another part and the farm as a whole and our income," he explains. "Because if you make a mistake, you lose money fast and hard."

Keeping everything working the way it is supposed to is a process that satisfies both his love of intellectual growth and his desire for character growth. "It causes you to grow, to be someone you weren't," he says. "You have times that you say to yourself, 'Two months before I would've done this, but now I realize some things I didn't then,' and so you see into yourself and the work more fully, and you make better decisions and develop your character."

These reflections take place around a long table in the kitchen, where our walks and talks with the Hales culminate. Eighteen-year-old Sarah, the oldest daughter, serves glasses of cold kombucha, a cultured tea she has made, to everyone gathered about, her parents, two younger brothers, a younger sister, and us. Sarah's newest venture is learning to make cheese, something that taking care of the milk cows has prompted, and she picks up the thread of Mike's observations about the challenges and rewards of the farming life. "The learning curves are always bigger than you think at first," she says. Her tone is cheerful and so is the bright attentiveness in her brown eyes. "I'll start something new and I know I'm going to learn a lot but really you have no clue how much. It just keeps happening, you just keep learning!"

"Whole new levels keep opening up," Mike agrees. He pauses and laughs. "Of course, there's plenty of setbacks, but still, it's nice to reach the point where you can tell the difference between having five years of experience and having one year of experience five times."

Asked if they regret any of the sacrifices they have made to do what they are doing, Mike smiles and says, "Not yet." Laughing, Sarah adds, "We'll see if it lasts!" She and Connie note that their last vacation was four years ago, when their commercial production was small and Mike was in his final year of teaching. With friends taking care of the animals, the Hales rented a beach house in Galveston for a week. Connie hopes that in the not-too-distant future they will be able to vacation more often. "Just to get away for a little while," she says, "and take a break from the fast pace of everything here." But Sarah remembers that the Galveston vacation came after five years without one. "So we weren't used to regular vacations even before we started farming," she points out. Then she makes an observation that demonstrates that, like her father, she too has read Salatin and taken him to heart. "Just living out on land, according to Salatin, makes you less and less interested in entertainment, in what most people think of as entertainment, anyway," she notes, and mentions times when she and the other children have chased straying sheep across the pastures. "Now that's entertaining!" she says, to the concurring murmurs and giggles of her younger siblings.

Connie and Mike laugh, too. "The sheep almost always find my sweet potato patch," Connie says, "and they'll devour it in no time. It's happened often enough now that I think of sweet potatoes just as a trap crop for the sheep. At least with sweet potatoes growing, we know where to find the sheep!"

"We now have a saying for these escapades, a little joke," Mike explains. "'When the Lord called us the sheep in his pasture,' we tell each other, 'going astray wasn't what he had in mind!'"

"Isn't that horrible?" Sarah says, and everybody laughs all over again.

Contact information

Mike and Connie Hale and Family
Windy Meadows Farm
8045 CR 4209
Campbell, TX 75422

Telephone: 903-886-7723
E-mail: mhale@worldlogon.com
Web: www.windymeadowschicken.com

Robert and Nancy Hutchins and Family
Rehoboth Ranch
2238 CR 1081
Greenville, TX 75401

Telephone: 903-450-8145
E-mail: hutchins@rehobothranch.com
Web: www.rehobothranch.com

Ross Farm

Learning to Listen to the Whispers of Nature

Betsy Ross was born in 1937 and grew up on Ross Ranch, a 10,000-acre goat and cattle ranch. It was established by her grandparents during the 1890s near Sonora, in west-central Texas. The closest neighbor was nine miles away. The place seemed so big, so endless, that when Betsy and her older brother and sister rode the land on horseback with their father, checking on their herds and grasses, Betsy sometimes worried that the others might crest a hill way ahead of her and she would be lost forever. The land she lives on and works now, the 530-acre Ross Farm, about sixty miles northeast of Austin, seems "little bitty" by contrast. "Really," Betsy says, "five hundred acres is nothing." But the work she does here—raising beef cattle on pasture and making liquid-compost extract—would challenge a person far younger than Betsy, age seventy at the time of our visit.

She generally maintains a total of 150 beef steers in five herds at any given time and occasionally also a herd of up to seventy cows and calves. Because she uses intensive rotational grazing methods, Ross Farm is divided into just over a hundred paddocks that range in size from three to ten acres, with most being five acres. Eight miles of underground PVC pipes deliver water from the farm's one well to sixty-gallon tubs placed in each paddock. When electricity fails and the well water can't be pumped, the cattle are moved to the farm's four stock tanks until power is restored.

Betsy's seventy-eight-year-old sister, Kathryn (born in 1929) lives at Ross Farm with Betsy. She moves the cattle to fresh paddocks every day, driving her Gator, a small, green all-terrain vehicle, to a given area and then walking the cattle to new pasture. Kathryn's gentle, soft-spoken manner lends itself naturally to the low-stress handling of livestock that they consider essential to animal well-being, and so does her expert attentiveness to animal habits. Growing up at Ross Ranch, she learned the ways not only of domesticated animals but also of wild ones. During World War II, when fur was needed to line the jackets of bomber crews, Kathryn trapped ringtails, raccoons, and foxes and sold the hides to the military.

Kathryn joined Betsy at Ross Farm in 2000, having lived most of her adult life in the New Orleans area after graduating from Louisiana State University with a degree in geology. Married to a geologist, she could not get a job with a big oil company, as her husband did, because of nepotism rules. But eventually she found a job teaching science in the private school their three children attended. "So I did get to use my degree somewhat," she says, "and since tuition was waived for faculty children, I put my kids through a good prep school almost for free."

Kathryn, Betsy, Kim, J. R., and Lil Bit resting in front of the farmhouse, Ross Farm.

The cattle on Ross Farm are a mix of Red Devons, South Devons, Red Angus, and Black Angus, medium-size breeds well suited to grass finishing. They come from Ross Ranch, which Betsy and Kathryn's seventy-two-year-old brother, Joe David (born in 1937), manages with his son. Joe David is an alum of Texas A&M, a lifelong rancher still active in national and international agricultural associations and a recently retired veterinarian. "Joe David works on the genetics we need and manages the breeding and calving," Betsy explains. "Between the Sonora ranch and this one, we own our cattle from birth to death. We know where they've been every day of their lives and what they've had in their mouths every day of their lives."

Betsy leads cattle from a paddock at Ross Farm.

The steers are twenty-four to twenty-eight months old before Betsy considers them finished and ready to harvest, considerably longer and more costly than the fourteen to sixteen months for conventional, grain-finished beef. "The shape I look for is something like a pound of butter with four little legs," Betsy explains. This is when the steers are just heavy enough to yield substantial cuts and reasonable profits, and just fat enough to yield meat that is tender and at peak nutritional value: lower in saturated fats and higher in beneficial omega-3 fatty acids than in the detrimental omega-6 fatty acids. Producing "nutrient-dense meat" is one of Betsy's passions. "I could probably make more money finishing them somewhat sooner," Betsy says, "but there's ethics as well as profits to think of."

Betsy became a commercial producer of grass-fed beef in 2004. The meat is processed at Readfield's, a state-inspected facility in Bryan, and marketed under the label "Betsy Ross Grassfed Beef." Though not certified organic, the Rosses use organic methods. A total of about 60 percent of the beef is sold to other pasture-based meat producers who sometimes supplement their own production in order to meet customer demand. Another 10 percent is sold wholesale to the Whole Foods store in Austin, and the remaining 30 percent is sold to people who come to the farm or purchase it through the farm website or from the People's Pharmacy in Austin and Old Thyme Gardens in Georgetown.

The liquid-compost extract business that Betsy also owns and operates, out of a metal barn at Ross Farm, is called Sustainable Growth Texas, LLC. Created in 2004, it is a soil-building business that stems directly from Betsy's schooling herself in soil biology as the basis for raising healthy animals and producing nutritious meat. Assisted by her son and daughter-in-law, J. R. and Kim Builta, and a team of five additional employees, Betsy makes compost, extracts beneficial soil microbes from it, suspends them in water, and sprays the solution on her own pastures and those of her clients. Like applications of compost, but more economical on a large scale, liquid-compost extract feeds soil organisms and stimulates their growth, making for healthy soil that can deliver more nutrients all the way up the food chain.

Ross Farm is the site of frequent field trips arranged by agricultural organizations for people who, like Betsy, are pasture-based meat producers or are thinking about it. Standing among a group of thirty people in a paddock of young steers and watching Betsy—trim, agile, and quick as a flash—climb to the top bar of a metal gate and straddle it so that she can see and hear everyone and be seen and heard herself, you might guess from how gracefully she moves and how engagingly she explains things that you are looking at not just a born rancher but an athlete and teacher, too. Those were my impressions in that situation, and then again on the cool, damp April day of our book visit, as Betsy hiked us through dense pastures of knee-high grasses and never got winded. Nor did she falter or need support when she stooped to her knees and spaded up some soil—to show and tell us about earthworms and nitrogen-fixing nodules on plant roots—and then just as quickly and smoothly stood back up again and led us on.

So it came as no surprise to learn, in the course of our conversation, that indeed Betsy was an athlete and a teacher. She played tennis and basketball from an early age, and in high school she played for three years on the basketball team, a team that won the state championship two of those years. At the University of Texas at Austin, she majored in physical education, and her first job was teaching physical education and coaching girls' sports in the Shiner, Texas, high school.

Sports metaphors come easily to her, including thinking in terms of teams. Steers in the final stage of finishing, for example, she likens to a college football team: "Once they weigh 950 to a thousand pounds, they're bulking up just like the best college football team. They're becoming fully developed, hitting their finest hour," she says. "But if we let them bulk up too long, it's like a team when the playing season's over. They start drinking too much beer and adding a lot of fat, extra fat, fat they don't need, and that's not what we want. We harvest them before that point, before they start adding extra fat between their muscles and skin. We want just a small layer of fat, for nutrition and tenderness. So, based on age and weight and eyeballing the size of the brisket and such, we pick them out for harvest when they're at the top of their game and are just fat enough but not too fat."

And, as many teachers and coaches do, Betsy also thinks in terms of realizing potential, not only in regard to people but to the soil and grasses and animals, too. "We do not name our animals but we respect them, each of them, all of them. We think their purpose in life is for us to treat them well, provide a good environment for them so they can fully express themselves and reach their potential and be really good food for us, to help us reach our potential," she explains. "But it's not that we don't have feelings for them, you know, and the older I get, the harder I watch that. I mean, each one is a life and I'm not so sure that in the scheme of things I'm more important than a beef is. I really doubt I am. Life is a circle, and we've kind of got it mixed up that we're at the top of the pyramid. We're just a piece in the circle, and the potential of all the pieces, including the life in the soil and the grasses, we should manage to reach their best potential, and be respectful." She pauses, then adds, chuckling, "I'm kind of out there a little bit as I get older."

Much of Betsy's adult life took her away from ranching. After teaching and coaching in Shiner for two years, she taught for a year in Abilene. In 1962, she married and became full-time homemaker, giving birth to J. R. in 1963 and a daughter, Susan, in 1966. Her husband's work as a newspaper editor and subsequently a lawyer took the family to Comfort, Kerrville, Austin, and Lampasas. When their marriage ended in divorce in 1977, Betsy returned to Austin and completed an MBA at UT in 1980. Drawing on her athletic past, Betsy wrote her thesis on marketing women's basketball to season ticket holders. She describes the program as "one of the neatest happenings of my life," a gift of affirmative action that opened doors to women and minorities. "It changed me forever," she says. "It set me up to do things I never would've done."

She remained in Austin and went into commercial real estate, owning and operating a firm from 1980 to 1989, when the market fell drastically and she had to close her business. After leaving real estate, she worked for the Texas Department of Insurance until 1999. Ann Richards was governor when Betsy entered the agency, as an assistant to the commissioner, and Betsy credits Richards with creating opportunities as enjoyable and challenging as those she experienced in business school.

"Richards opened state jobs up for a lot of women and a lot of minority people, and I was lucky to be one of them," Betsy says. "She was an outstanding example of what one person can make happen, of what any one of us in our own little ponds can do just because of the decisions we make." Long accustomed to thinking of herself as a feminist, Betsy found Richards to be a kindred spirit and one of several important role models. "Before I married, I'd done all my own stock portfolio trading, and then I married and had to have my husband's approval!" Betsy recalls. "And my attitude was, 'What are you talking about? I wasn't brought up that way! Not even my father did my trading!' But that was the law, till along came feminism. Betty Friedan. Barbara Jordan. And later, Ann Richards. Those people changed my life!"

After Richards left office, Betsy's job within the agency changed. "I kept getting demoted every time we had a new commissioner," she says. But she quickly came up with an idea for a project that would benefit the agency and be enjoyable for her as well—learning computer programming, creating a website for the agency, and teaching web skills to other employees. "Actually, I didn't know a lick of programming, but I knew the good old girl network," Betsy recounts, "and I found out who was doing what at the other agencies and got started." And then, as she had done before and has done since,

Kathryn on the Gator at Ross Farm.

Betsy built a team. "There were twelve of us at the insurance agency, and we made this commitment that we'd share everything we learned. And I did all the teaching and we ended up with about a hundred people from a number of agencies on our website-building team, including a bunch of men who were just wonderful guys. And before long we were relieving the insurance agency of at least eighty percent of what would've been phone calls. It was a nice learning experience. Working together as a team is so powerful, so powerful!"

In 1992, while still living in Austin and making the transition from real estate to the insurance department, Betsy began coming to Ross Farm on a regular basis. Joe David had purchased the farm in 1975, in partnership with his former college roommate, who was from the area and managed the place. They raised beef cattle conventionally on pasture, hay, feed corn, and grain and shipped them to feedlots. The partner's death left Joe David bereft of a friend and in need of a manager, so he asked Betsy to step in and check regularly on the cattle and pastures.

Compared with the thin soils, sparse grasses, and arid climate of Ross Ranch, Betsy found the claylike blackland soils, high rainfall, and fast-growing vegetation of

Ross Farm to be a new world. Plus, despite returning many times throughout the years to Ross Ranch for family gatherings, Betsy had not done any agricultural work since her youth. "I didn't know how to drive a tractor anymore," she says, "and the first time I got on one out here without someone coaching me, I forgot to pump the brakes, so the hydraulic fluid didn't go where it's supposed to, and the tractor was this live, wild thing, and I was going downhill and went through every fence! All the workers were just scattering and watching and I was yelling, 'Help, help!' And finally I kind of maneuvered it up a slope and brought it to a stop."

To avoid worse scrapes, and to get her own sense of the place and develop a sound management plan, she realized she needed the guidance of a real pro and a good teacher, and she turned to Joe David. Some weeks they talked by phone almost every morning, and occasionally he joined her during her stints at the farm. In addition, he urged Betsy to participate in workshops and conferences about grazing issues and other aspects of cattle management and meat production, and to bring experts to the farm for consultations, and she readily agreed. She went to programs hosted by Allan Nation, cattleman and editor and publisher of the monthly *Stockman Grass Farmer*, and attended conferences of Holistic

Betsy examines signs of healthy microbial life in the soil of Ross Farm.

Management International, an organization founded by Allan Savory, an internationally recognized rangeland expert whose work Joe David had studied since the seventies. She began reading *Acres USA*, a national publication devoted to "eco-agriculture," and attending conferences sponsored by the magazine. At the farm, she hosted rancher Walt Davis and other grazing experts associated with the Noble Foundation, a widely known sustainable agriculture organization based in southern Oklahoma.

"This long stream of people came to help me," Betsy recalls, "and then Joe David sent his forage man down and we became wonderful friends, and Joe David realized we could use this man's help regularly, to think things through. So for a while he came every weekend and we'd test every pasture and sit down and do a plan, a six-month plan or a three-month plan, and measure and grade what was out there. And then some folks from A&M started coming out and doing their sheet tests on poop, a test that predicts an animal's weight gain by the color and consistency of poop. We were studying all kinds of things, trying to figure out the best things to do."

Something Betsy did not need much help studying was grasses. She had learned to identify them as a young girl, first with her father and later through 4H projects and competitions. "Our daddy never let us ride out without our little packsaddles full of seed. He was always throwing out seed into bare spots and areas where growth was sparse, and he had us do it, too," Betsy explains. "So we always paid attention to grasses, along with whatever else we might've been seeing to."

At the age of fourteen, she became the first girl to make the state 4H grass identification team, which Joe David was also on. Betsy relishes the memory, and not just because their team won. "They had to unlock an entire dorm for me at A&M because I was the only girl, and Daddy had to go and stay with me there, and all the boys were trying for scholarships and if I won something, they'd say, 'That's really mine!'" she recalls. "And later—all your life, you know, you run into the people you grew up with in surrounding towns out there—and later they'd say, 'I remember you from grass judging!'"

Betsy also entered mohair judging and other sorts of wool judging, but she was best at grass identification and loved it the most. And as she returned to agriculture at Ross Farm, she found grasses and other forage plants as interesting as ever: the native bluestems, switchgrass, Indian grass, eastern gama grass, sideoats grama, and native clovers; and the nonnative Bermuda grasses, ryes, sorghums, and legumes like alfalfa, crimson clover, and hairy vetch. "That's what really captivated me here, right away, was all these grasses growing so fast!" she says. "In West Texas you have to wait for rain, wait a long time, and it'll take forty-five days out there for a grass to recover, whereas here they can recover in seven to ten days! It's just amazing!"

Within about three years of her reeducation into agriculture, Betsy was persuaded that intensive rotational grazing was the best method, and by 1995 she began implementing it at Ross Farm. Most of the people and programs she learned from emphasized the many health benefits—to soil, plants, animals, and people—of not grazing cattle on large tracts for long periods, as in conventional grazing, but concentrating them briefly in small areas and moving them frequently. Conventionally managed cattle both undergraze and overgraze a large area at once, selecting the plants they like most and eating them to extinction while ignoring plants they don't like. In intensive rotational systems, by contrast, animals concentrated in subdivided pastures, or paddocks, can't "pick and choose" so extensively and consequently graze and stimulate the growth of plants more uniformly. And in small paddocks, their wastes are more concentrated

Liquid-compost extract is loaded for transport from Ross Farm to a client's farm.

than in large pastures, which means they more uniformly and effectively feed the soil and plants.

Betsy was also discovering during this time that for weed control and for fertilizing, chemicals were not working. "Out in West Texas we never used chemicals much but here we did," she explains. "Whenever the grasses disappeared under the weeds, we'd put out some chemicals, put out some herbicides to beat back the weeds, and then we'd go back with synthetic chemical fertilizers to make the grasses grow. But the weeds just kept coming, and kept beating out the grasses. So we'd spray them some more and we'd try spot burning and pulling, but they just kept coming. And finally I said to Joe David, 'This chemical thing is just not working on these weeds. I've had it! I'm tired! We've pulled 'em, burned 'em, and nuked 'em, and it's time to do something different."

Joe David asked her what she thought she should do, and she told him she didn't know, but something, and he said, "Well, whatever you do, don't go organic." To which Betsy replied, "Well, what's organic? It's a growing system and not some tree hugger way of life, if that's what you're thinking."

In almost no time, Betsy went organic. As with many

people, for Betsy it was initially a matter of what she did *not* do more than what she did—the chemical fertilizers, herbicides, and pesticides she did not use on the pastures, the artificial growth hormones, antibiotics, and feed corn she did not use on the cattle. But by 1999, when she retired from work in Austin and moved to Ross Farm to live and work full time, she could see the inadequacy of this approach to organics. The soil, the grasses, and the cattle were not as healthy and productive as she wanted and figured they could be, with further thought and work on her part.

"If you keep asking yourself questions, teachers come to you," Betsy says, "and Bonnie and LeRoy Sladek, these great friends of ours who run the Old Thyme Gardens nursery in Georgetown, brought in a speaker in 2001 or '02 who talked about manure tea. And it made a lot of sense, and I bought a bunch and put it out and got some good results."

Around the same time, some USDA friends of Joe David gave them a copy of the *Soil Biology Primer*, a booklet published by the Soil and Water Conservation Society and written primarily by Dr. Elaine Ingham, an Oregon State University microbiologist. The primer describes the interrelated life cycles of soil organisms and how, when the soil is properly tended and nourished and these organisms are well fed, they enhance agricultural productivity as well as air and water quality. "I read it right away," Betsy recalls, "and said, 'Oh! This is too compelling! Why are we doing anything else? We need to learn everything we can about feeding all this biology in the soil!"

Ingham, as an outgrowth of her academic research and writing, maintains a website, www.soilfoodweb.com, and a business called Soil Foodweb, Inc. Both are intended to educate a broad public about the life in the soil and how to feed it. Compost tea is one of the main soil-feeding measures she advocates. Soil Foodweb trains people in this and other techniques and certifies them to be Soil Foodweb advisors, a credential they can use in operating their own biologically based soil-improvement businesses. Betsy went immediately from the *Soil Biology Primer* to the information on Ingham's website and studied it. After that, she read all Ingham's other books and then, with her son J. R., flew to Oregon to take certification classes.

The upshot was that Betsy and J. R. began making their own liquid-compost extract and spraying it on the pastures of Ross Farm. They watched for results and had everything laboratory tested for certain nutrient levels—the soil, the grasses, the cattle, and the meat—and found strong improvements at each level. They soon went commercial with their liquid-compost extract, by establishing Sustainable Growth Texas, LLC.

Their composting material comes from both Ross Farm and commercial sources and includes woody stuff, grass, alfalfa hay, fish, and manure. Big, black-brown mounds of it are clustered just outside the barn housing the small metal extractor that turns it into liquid fertilizer. The business has grown rapidly. Clients include not only farmers and ranchers, many of them in pasture-based meat production, but also growing numbers of urban retirees who are moving to rural areas and are not necessarily involved in any agricultural production.

"A lot of our retiree clients have these little twenty-five- or forty-acre places, and they want to make them pretty. They know they need to improve the soil and plant life to do that, and they don't want chemicals. Some of them want to restore native grasses, and we help them with that." Betsy says. "A lot of them just want to spend time with their grandkids out on their land, and they don't want their grandkids exposed to a bunch of chemicals."

As for weeds, hers or anybody else's, those vexing things that drove her away from chemicals and toward organics in the first place, Betsy has learned to see them in a new light—not necessarily as a problem but as a piece of nature's circle, a living thing to understand, to listen to, like all the other pieces. "Weeds are whispers from nature, and there are whispers all the time, all the time," Betsy explains. "Nature's saying, 'See what this weed or this insect is trying to tell you about your soil, about what you're doing or not doing that you might need to change.' Or maybe not change. I've learned to live with weeds and so have our cattle. A lot of weeds are palatable. So, whereas we used to look at our weeds and say, 'Oh isn't that trashy looking, isn't that terrible? We better go spray!' We now say, 'Well, isn't this interesting? What's Mother Nature trying to tell me here?'"

The main hub of activity at Ross Farm is a hilltop that overlooks pastures in every direction. This is where the metal barns and sheds are, the compost piles, the trucks with spray rigs, a travel trailer, and the white frame house with blue trim where Betsy and Kathryn live. This is where you hear diesel engines rasping along, cell phones going off, and people calling out to each other. But south of the hill and along the San Gabriel River, another part of the farm is a haven from commotion, a place prized for its broad river shelf and enormous old bur oaks. "It's called the Tabernacle," Betsy says, as we walk around the area with her and Kathryn. "In the early 1900s, through the twenties and thirties, people came here on their horses and buckboards and camped for prayer meetings and singings."

"And way before that," Kathryn adds, "the Indians used this place."

Betsy points out a low-growing vine a few steps in front of us. It has lobed leaves and small, purple bell-shaped flowers. "This is rare," she says. "What is it, Kathryn?"

"Purple leather flower vine," Kathryn answers. Much shorter than Betsy, and wearing a red baseball hat with the bill set low on her brow, she has to tilt her head up to meet her sister's eyes. Betsy thanks her. She's been helping Betsy with names since the day Betsy was born—when Kathryn came home from school with a lesson about Betsy Ross and the flag on her mind and was allowed to name the new baby.

Kathryn and Betsy are pleased to see the blooming vine, and not just because of its beauty. It is a sign of revived life in the soil and of the sisters' victory over the county road crew. "We stay after those folks, and it looks like we've maybe finally talked them into quitting herbiciding here," Betsy says. "And we keep putting out alfalfa hay here, too, as a mulch to feed and heal this land."

Getting things in good condition for the future and teaching the younger generation of their family all that the older is learning about the soil, intensive rotational grazing, and grass-fed meat production are increasingly important to Betsy and Kathryn. J. R. and Kim are by far the most involved of the younger generation. Betsy and Kathryn believe they will move the farm forward along the lines Betsy is still charting.

"What we're doing here isn't what you read in the text books," Betsy says, "and it's not what the ag agents are going to tell you. It would be a shame to acquire all this knowledge we've been acquiring and have it go to waste. There's all the day-by-day work and all the thinking ahead. We're always thinking six months down the road. What do we have to do today to get the grass ready six months from now? When we oversow one pasture in a certain forage, say a warm-season forage, then we have to think ahead and be ready to clean it off when it's time to plant a cool-season forage. Well, somebody's got to

The "Tabernacle" along the San Gabriel River at Ross Farm.

do all that thinking and planning, and this year, for the first time, J. R. did it, and I thought, oh my, he's going to really stump his toe but that's the way you learn sometimes. And it turned out he didn't stump his toe, not this year anyway!"

Yet, however urgently the future pulls them forward, some days they can hardly believe that they're the passing generation, not the upcoming one. And in fact they have been the oldest generation in their family for barely three years, only since their mother died, at the age of ninety-nine. "At our age," Betsy says, "three years is like the blink of an eye, and we had her so long, we haven't quite realized she's not with us anymore."

"Still feel like orphans," Kathryn says.

"We do," Betsy agrees. "And sometimes we just look at each other and say, 'We're only kids! What are we doing running things?'"

Contact information

Betsy Ross
Ross Farm
900 CR 493
Granger, TX 76530

Telephone: 512-862-3240
E-mail: betsy@rossfarm.com
Web: www.rossfarm.com

Dairy and Cheese

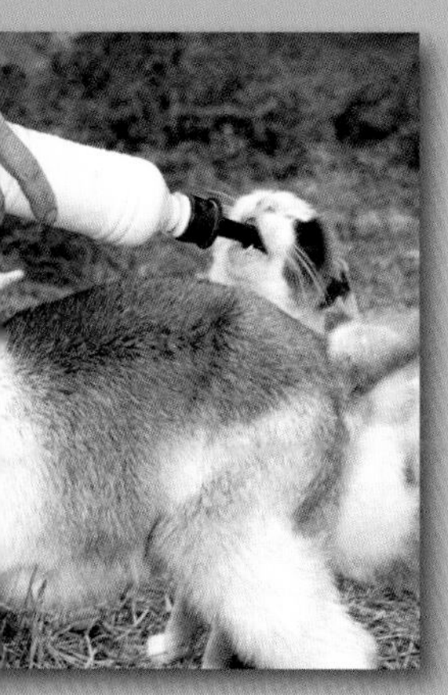

Pure Luck Farm and Dairy

From the First Generation to the Second

Combining animal husbandry with crop production in a mutually enriching way is often invoked as an ideal form of sustainable and organic agriculture. Yet such mixed, or diversified, farming is seldom done commercially. Most farmers base their commercial production either on plants or on animals, some because they are intellectually and emotionally drawn more to one than the other, and some because encompassing both might pose insurmountable labor and marketing challenges. The Boltons and Sweethardts of Pure Luck are exceptions to this general rule and run one of the most diversified farms in the state. They operate a Grade A goat dairy and produce award-winning cheeses and also grow flowers, vegetables, and certified organic herbs. Located about six miles west of Dripping Springs and about thirty-five miles west of Austin, Pure Luck consists of a ten-acre parcel and a fifty-acre parcel, separated by a narrow road.

The ten-acre segment lies along Barton Creek, which is visible from any distance only as a fringe of trees, mainly live oak and cedar. Up from the creek and near the road, the same types of trees, along with hackberries and cedar elms, form shady clusters around a limestone cabin-turned-office, a small cedar-sided farm stand and walk-in cooler, and the Bolton family's white frame residence. The acreage is mostly open, however, and here, in twenty-five plots of loamy, creek-bottom soil, the Boltons and Sweethardts grow a seasonally changing variety of market crops and cover crops. Each plot is 70 feet wide by 100 feet long, for a total of eight cultivated acres. Five acres are dedicated to herbs and are certified organic, while the remaining three acres, dedicated to flowers and vegetables, are not certified. Certification costs are based on acreage, and though the Boltons and Sweethardts use organic methods with everything they grow, they find that certification gives them a significant market advantage only with herbs.

The fifty-acre segment across the road is studded throughout with live oaks and cedars, providing shade and browse for just over a hundred mostly Alpine goats, plus a few Nubians and Alpine-Nubian crosses. The goats' diet consists less of browse, however, than organic whole oats, sunflower seeds, grains, alfalfa, and whey, so the goats don't have the run of the entire acreage. Rather, they roam and frisk on several fenced acres fronting the road. Inside this enclosure is a large cedar shed, and under its long, wide roof the goats take shelter whenever they want. Their manure is scooped frequently and mixed into a big compost pile outside the enclosure, which in time fertilizes the soil of the horticultural plots. A short distance beyond the goat yard is the dairy barn, a compact, red steel building that houses the milking and cheese-making facilities. And

Sara and Denny Bolton on the back porch of the Pure Luck farmhouse.

not far from the dairy barn is a white clapboard bungalow. Amelia Sweethardt, an owner of Pure Luck, lives in it with her husband, Ben Guyton. Amelia is the second daughter of Sara Bolton, who founded the farm, and is among the second generation of her family to work here.

Sara died in November 2005, at the age of fifty-two, of breast cancer. In the months before, she led her family in establishing the farm as a corporation. Sara's husband and business partner, Denny Bolton, is president, Amelia is vice president, and her oldest daughter, Gitana Sweethardt, is secretary-treasurer. (Sara's youngest daughters, Claire and Hope Bolton, minors at the time of incorporation, assumed no formal role.) Denny is in charge of horticulture. Amelia manages the dairy and cheese making and oversees the day-to-day operations of every aspect of Pure Luck. Business administration is shared by Denny, Amelia, and Gitana.

We visited Pure Luck twice, six months before Sara's death and then again seven months after it. During the first visit, we spoke mainly with Sara and Denny; during the second, with Amelia and Hope.

It's Sunday morning in early April. The weather is cool and drizzly, and a wire fence near Sara and Denny's house holds many clothespins, but no clothes. "On a prettier day, you probably would've seen our laundry," Sara says, and explains that just a little earlier she and Denny and Hope and Claire rushed out to retrieve the things they had hung the day before. They have always loved sun-dried clothes and sheets, yet several years ago they bought a dryer. It worked fine until last year but then broke and didn't get fixed; the family realized how much they had missed the scent of the sun in their clothes. "And besides," Sara says, "sunshine's free! I don't know why we ever got a dryer!"

Sara has wavy, gray-brown hair, and she is small and agile, slipping easily between wire fence strands or crouching to bottle-feed a kid and—just for grins—even a pet cat. Denny is tall and stout but agile, too, and though less of a talker than Sara he is attentive and friendly. His curly hair, round face, and big smile give him a youthful look that belies his sixty-three years and the eleven years he has on Sara. As Sara and Denny show us around the farm, a big white dog named Cooper joins us now and then. He's a feisty, three-year-old Great Pyrenees. An older one, Queso, checks us out just once, then goes his own way.

The story of Pure Luck is, in the beginning, Sara's story. Born in Boston but raised in northern California, the daughter of a Palo Alto architect and granddaughter of a San Francisco oil magnate, Sara arrived in Austin in 1974, with no plans to stay. She had not even planned to go to Austin in the first place. She was twenty-one and it happened on a whim. "I'd been traveling in Mexico for a while and was headed home," Sara recounts, "but I sort of took a right fork instead of a left fork and somehow got to Austin." Before long, she was putting down roots there, human roots. In January 1975 she gave birth to Gitana, and in May 1977 to Amelia. Still, California felt like home—her parents, her younger brother and older sister, aunts, uncles, and many cousins lived there—and Sara might have gone back. "But Gitana and Amelia's dad and I split up when they were very young," she says, "and I didn't want to separate the girls from their father, so I stayed in the area."

In 1979, Sara purchased the ten acres along Barton Creek. Previously a tomato farm, the place had several outbuildings, a simple but adequate house, good soil, and some nice swimming holes. Growing up, Sara had spent a lot of time at her grandfather's Sonoma County ranch, and she always treasured the space and freedom to explore, work with animals, and grow things on a scale you couldn't in a town or city. Ten acres seemed big enough to let her enjoy something of that freedom again and share it with her daughters yet small enough to be physically and economically manageable. She was used to keeping day-to-day expenses low and working only odd jobs. "I just didn't spend much money," she says. "I worked at a plant nursery, I drove a school bus, stuff like that."

She was also into organic vegetable gardening, and as she and the girls settled onto the farm she grew more of their food than before. And she had just enough experience with goats, from tending to a few for a vacationing friend, to know that she really enjoyed their playfulness and fresh milk. So she began keeping a few Alpines, high-volume milk producers whose short hair grows in various patterns of brown, black, white, and gray. Increasingly fascinated, she got interested in particular bloodlines and soon developed a small herd she wanted to register with the American Dairy Goat Association. The process required her to pick a herd name, and for reasons she can't remember, except that she must have

Sara Bolton feeds a young goat and farm cat.

been feeling lucky, living on a fine piece of land with her two little girls and her milk goats, she called this herd "Pure Luck."

"Almost everyone who raises goats does it for the love of the animal, just loves their personalities, and takes really good care of them," Sara says. From the first, she considered a healthful diet the key to a healthy herd, and she fed hers only whole, organic grains and alfalfa and used antibiotics and pharmaceutical wormers only in life-threatening situations. "I pamper my goats, and it's so much fun. I mean, just look!" She pauses and points to a couple of kids pouncing on some hay bales, and to another pair butting heads. The kids amuse Denny also. "Pranksters," he observes, "goats are great pranksters." "Such fun to watch and take care of," Sara continues, "which is what got me going, that and the milk for the family, so good and fresh."

About three years after Sara purchased the first ten acres of Pure Luck, as the farm itself quickly came to be called, the fifty-acre parcel across the road went on the market, and Sara seized the opportunity to buy it. A firm believer that land is one of the best investments a person can make, she figured the value would only appreciate and, among other possibilities, come in handy should she need to sell it for her daughters' education.

Sara and Denny met in 1983. He owned and operated a plant nursery and landscape design business in the area, and Sara's work for another nursery brought them into contact. Born in Omaha in 1942, Denny, along with his two younger brothers and an older sister, grew up living in several places, wherever their father's air force service took them. One was Austin, and Denny really liked it, but they soon moved on and he graduated from high school in Little Rock, Arkansas. After that he did his own moving around, at one point spending a couple of years in Houston. By 1970, though, Austin drew him back. Alternative culture was thriving, and he liked the general scene. "Sara and I are kind of from the hippie generation," Denny says, "and it felt comfortable around here." In 1984, a year after meeting, they married, and horticulture at Pure Luck expanded way beyond Sara's family-size vegetable garden. Denny created growing plots, greenhouses, and a drip irrigation system, and he and Sara went into the landscaping business together. The size of their family also expanded, with the births of two daughters, Claire in 1986, Hope in 1989.

During the late eighties, Sara and Denny switched from growing landscaping plants to food crops and were among the first in the state to receive organic certification through TDA. They also were among the first to help establish what later became TOFGA. To get started in vegetables, Denny and Sara planted about half an acre of cucumbers and sold the entire crop to the local Whole Foods. The next year, they planted at least twice as many cucumbers, expecting that once again Whole Foods would buy them. But Whole Foods had begun buying less on a store-by-store basis than through its own regional purchasing-and-distribution arm (then called Texas Health Distributors, and since called Southwest Distributors), and other farmers were already supplying the company with all the cucumbers it needed. With no time to waste, Sara and Denny went knocking on other doors, and luckily they were able to sell most of their cucumbers to several small natural food stores.

"Denny and I hardly ever think things through," Sara says, noting that they faulted themselves for their predicament, not Whole Foods. "We do something because we're interested in doing it, and then figure out what comes next!"

Their cucumber experience made them wary of growing not just more cucumbers but growing vegetables for their primary crops, period. There appeared to be enough other organic farmers selling vegetables wholesale that establishing as reliable a local niche as they wanted might be impossible. Sara wondered if they could find a readier local market for fresh-cut, organic culinary herbs. The idea instantly appealed to her and Denny. Yet, with their cucumber scrape still vivid, they decided not to rush out and plant but to think the move through as thoroughly as possible. "We scoped out health food stores, grocery stores, and restaurants and saw that no local organic growers were raising culinary herbs and supplying them in fresh-cut bunches," Denny explains. "So we went for it. We selected certain herbs and started growing several varieties of each—basil, thyme, mint, parsley, rosemary, sage, oregano, marjoram, dill, and some others—and this became our first real niche."

H-E-B and Whole Foods signed on as their biggest clients, though smaller stores, including the Wheatsville Co-op, bought from them, too. The market proved such a good one that they decided to enlarge the herb production in some complementary way, and in 1990 they added flowers to their crops, mainly zinnias, sunflowers, and marigolds. By 1996, fresh-cut bouquets provided about half the family's income, with fresh-cut herbs and a limited selection of vegetables—a few varieties of spinach, lettuce, and summer squash—providing the other half. About 1997, they added retail venues to their wholesale markets, starting with the Sunset Valley Farmers' Market (originally the Westlake Farmers' Market) and a stand at their farm, plus the Austin Farmers' Market in 2003. On market days, their daughters helped them sell.

Sara did not consider making cheese commercially until some point late in 1994. By then, she had been making cheese for their family for years, and so routinely that it had become a staple in their diet. She started with a few basic recipes. "I'm an experiential learner, and it's pretty basic stuff anyway," she explains. "It's just not that hard, especially if you have good goats and good milk and do it often, you know, stay in practice. And if you don't mind washing vessels and utensils a lot and keeping things really clean." She notes that the dishwashing aspect cannot be overemphasized. "Some people have the idea that cheese making is kind of glamorous," she says, "and it's just not. Interesting and satisfying? Fun? Yes! Glamorous? I don't think so!"

Though Sara never had any pretensions about cheese making, she could see that she would become good at it. She branched out from simple chevre and feta to more complex ripened cheeses, and the taste and texture of each type only got better. Family and friends began asking why she didn't turn cheese making into a business. The idea grew on Sara, and she decided that, as long as they kept their commercial cheese production small, it would not take much more effort than she was already making. And, given their thriving market in herbs and flowers, it seemed like an almost ready-made opportunity to add a high-premium product to their offerings and not break stride. Denny could deliver cheese to Whole Foods and H-E-B on the rounds he was already driving, and they could sell cheese at farmers' markets along with everything else. So in 1995 they constructed the small, red dairy barn and got state licensing to operate a Grade A goat dairy. Sara started mainly with chevre and feta, and it sold very well, much better than they anticipated. Whole Foods and H-E-B, for marketing purposes of their own, wanted greater volume as soon as possible.

"As usual," Sara says, "we realized we'd jumped in without thinking things through."

"Did we ever," Denny says. "We thought we could milk just ten or twelve goats, something like that, and develop a cheese niche just on that!"

But, in fact, they soon were milking about seventy goats, out of a herd of over a hundred, numbers they have more or less maintained, they explain, laughing with each other, and us, as they reminisce. These numbers meant that they had to add on to the dairy barn almost immediately. Even so, the whole building is only 12 by 55 feet, the sum of the three rooms that make it up: a cheese plant (12 by 28 feet), where milk is pasteurized and cheese is made; a milk room (12 by 12 feet), where milk is stored in a three-hundred gallon holding tank; and a milking parlor (15 by 12 feet), where eight goats at a time are held in stanchions and manually cleaned, then mechanically milked two at a time. Sara explains that the dimensions of all three rooms are tight for the volume of cheese they produce, and that the limited facilities in the milking parlor make milking a time-consuming process: three and a half hours each morning and each evening.

"We should've added more space while we were adding," Sara says, "but we didn't think through that deal, either! And we didn't even think to have a walk-in cooler, not at first. We had an extra fridge we'd bought used, was all. And even crammed, it couldn't hold everything we were making."

Amelia Sweethardt covers rounds of chevre in the Pure Luck cheese room.

"And really," Denny says, "we built the whole dairy totally backwards. We have to pass through the goat yard to get to the milking parlor. The goat yard should be, oh, like right behind the milking parlor."

"Shhh!" Sara tells him, joking, and nods at us. "They're recording all this!"

"Well," Denny adds, chuckling, "we're not the only ones to make it backwards. Who was it, Sara? Which friend said, 'You build one dairy and then the next one you know what to do'?"

In 1998, only three years into commercial cheese making, Sara entered the American Cheese Society's annual competition and won four awards. There were 320 entries that year from seventy-seven artisan producers, and Sara's chevre placed second, both in the general competition for chevre and in the category limited to farmstead chevre. A ripened Camembert-type cheese called "Del Cielo"—after Joe Ely's recording "Gallo del Cielo," a ballad about a fighting cock—placed third in the general competition for this type of cheese and first in the farmstead category.

The farmstead category is limited to cheese made by

people who operate their own dairy and produce their own milk, bringing an elite group of cheese makers into competition with each other. Most cheese makers, even artisan producers, buy their milk. "Dairy and cheese making are two separate businesses, and it's really hard to do both." Sara says. "Sane people don't!"

But Sara loved goats before she loved cheese making, and with Amelia's help she developed the dual operation into Pure Luck's most remunerative enterprise. To the chevre, feta, and Del Cielo, Sara added several more cheeses: Sainte Maure, a traditional French, surface-ripened one; Champagne Brunch, a brie-like creation; Hopelessly Bleu, a blue named for Hope; and Claire de Lune, a semifirm, ripened cheese named for Claire. She gradually increased the production volume of each, though chevre and feta still make up the bulk, and they now produce an average of 370 pounds of cheese each week. Cheese sales have accounted for 60 percent of the farm's annual income since around 2000, with cut herbs bringing in about 20 percent and flowers combined with vegetables also 20 percent. And every year at the American Cheese Society competition, now the largest cheese competition in the country, Pure Luck continues to win awards (a total of twenty-three when we visited Sara, and twenty-seven when we later visited Amelia), bringing national distinction to the farm.

Sara and Denny show us into the cheese plant—a small room with stainless steel counters, sinks, and tanks—as Amelia opens it up for work. Sara directs us to step into a pan of disinfectant before crossing the threshold and tells Amelia right away that everybody's done the "foot thing." Amelia nods her approval and doesn't demur when Sara adds, "Amelia's our boss!" Amelia is wearing a white apron over a green T-shirt and pants and has covered her light brown hair with a white net. She is welcoming but all business. This is day three in a four-day cycle of chevre. From a tall rack of white trays, Amelia slides one out, revealing two dozen small round basketsful of curds. They are in the final stage of being drained of whey. "Tomorrow morning," Amelia explains, "we'll roll some of the chevre in herbs, leave some plain, and package all of it in containers, lid and label them, and put them in the walk-in cooler for Tuesday delivery."

Almost everything in the cheese room is white—the trays, the cheese, the walls—and with windows on three sides the room is replete with light even on an overcast morning. "It's a beautiful environment for work," Sara says. "There's always a nice light, and you can look out at the goats and the grass and the trees and the sky."

It is not a time for us to linger, though. Amelia has work to do, plus something fun to watch is happening outside. Amelia has turned on a spigot that delivers whey into a galvanized tin trough. Hearing the whey and smelling it, the goats trot from the shed and take impatient turns lapping it up.

As demanding as their diversified farm work is, Sara and Denny say they have never felt deprived of things they wanted or any more tied down than most people running their own business do. They have been able for some time to afford medical insurance, for example, but they prefer not to purchase it and instead focus on exercise and nutrition as keys to health. And they have always been able to take vacations, leaving things in the hands of four long-term employees and eventually also Amelia, and more recently Gitana as well. Gitana earned a degree in microbiology from UT Austin, married, and worked in a state office until the birth of her son in late 2004. Soon after that, she agreed to work part time at Pure Luck, in accounting. Gitana's husband works in the governor's office, in a department concerned with small-business development, and Gitana had the option of being a stay-at-home mom, Sara explains, and took it. "Bringing baby Will out here to work's the same thing," Sara says, "and it's fun for all of us."

In addition to taking vacations, Sara and Denny have also had the leisure to pursue interests unrelated to farming. Denny writes mystery novels and published one called *Hippie Hollow: Murder on a Nude Beach*. Sara and fifteen-year-old Hope take fiddling lessons together, from a woman who plays in an Irish rock band in Dallas and comes to the Austin area once a month. "Hope, of course, makes progress much more quickly than I do," Sara says proudly. "She could be awesome but she needs a lot of encouragement to practice and that's why we take lessons."

The main hardship they have experienced as farmers is that neither is mechanically inclined. This has posed more problems with the dairy and cheese making than with their horticulture. "We're really isolated here as far as other goat dairies or cheese producers or repair people for dairy equipment," Sara says. "So repairs for the pasteurizer or the bulk milk tank can take a while and cost a lot of money, and sometimes cause emergencies or just huge inconveniences."

But the pleasures of the farming life so overwhelm any minor pains that Sara and Denny envision not just Pure Luck in their future but also a second farm. In 2003, with owner financing, they began purchasing an eighteen-acre farm located a hundred miles north of Sacramento, on the western edge of California's Central Valley and the eastern edge of the Mendocino National Forest. "For a long time, we've wanted to live in California," Sara says, "and we thought, well, here we are getting older, let's at least get headed in that direction." Fifteen acres of the farm are irrigated pasture, the other three orange groves. "And," Sara emphasizes, "it has one of these big, old beautiful California barns that I think are so wonderful, about four thousand square feet. And it's all fine, solid wood. Very, very nice. Incredible, really."

Denny is equally enthusiastic. "We fell in love with the barn first," he says, "and the house is all right, too. We're not sure how old, but we know it was built somewhere between the 1890s to the 1920s. I can see myself there, writing and watching the orange trees grow."

"And I," Sara says, "can see myself dairying."

As usual when they start something new, they are not sure exactly how it will work out. "We're trying to formulate a plan," Sara says. "Trying! Which is more than we've often done, but we've got to. We don't really have any savings, so that makes just going out there and starting up something of a problem."

Whatever happens, though, they cannot picture selling Pure Luck. They briefly flirted with the idea, but that's as far as it went. They might divide their time between California and Texas, but no matter what, Pure Luck will always be home. "We're thinking we can slowly set up out there and the older kids can run things here," Sara says. "We can borrow to start up out there."

Meanwhile, Sara, far from vacating Pure Luck even mentally, is thinking about making the dairy less grain based and more pasture based. She and Amelia are talking about its feasibility. Do they have enough land to create pastures for rotating just over a hundred goats? Or would they have to reduce the herd size? But if they did that, what would it mean for cheese production? To think about these and other things, Sara goes "down in the green a lot," she says, by which she means the big patch of cover crops where we walk as our visit comes to an end. "I used to do a lot of the tilling down here," Sara reflects, "and Hope, when she was little, would follow behind the tractor, and later she would say, 'I love the smell of dirt!'" Now lush in oats and wheat, Austrian winter peas, and brilliant heads of crimson clover, this

Cover crop of oats, wheat, winter peas, and clover at Pure Luck.

green manure soon will be tilled under and planted in basil and other warm-weather herbs. Sara picks a pea pod, eats it, then picks pods for Denny and us. "Crisp, juicy, sweet little things, aren't they?" she asks. "Such good grazing here," she adds. "What's not to love?"

It's another Sunday at Pure Luck, this one in late June fourteen months after our first visit. A little past noon, it's mostly clear and still and plenty hot, and even for a Sunday unusually quiet. On the whole farm, Amelia and sixteen-year-old Hope are the only ones home, and they just got back themselves. They spent the morning at a brunch and baby shower and are still dressed for it. Amelia is wearing white linen pants, a yellow linen top, and a white, broad-brimmed hat. Hope is wearing a ruffled, floral print skirt and a lacy white top and has clasped her hair into a topknot. The shower was for one of their employees and his wife. "Their baby will be born next month," Amelia says. She gestures at a stroller and some toys on the back porch of Hope and Denny's house, where our conversation begins. "These are

Gitana's baby's," she explains. "We're all set up. This new baby will fit right in."

Amelia, twenty-nine and less than a year into her new role, describes herself wryly as "el jefe." Yet she emphasizes that, although she oversees all the day-to-day operations, she is just one part of a team. "A really good team," she says. "Denny and my three sisters, my husband, Ben, and our four employees. And I think we all have a huge sense of the value of this farm and what we do here. I mean, our extended family and our employees are all supported by this place!"

Amelia reports that since Sara's death they have stopped selling retail and so no longer maintain a farm stand or participate in farmers' markets. "We just don't need them anymore," she explains. But they have made no changes in production or day-to-day workflow. Cheese production continues to average about 370 pounds a week, with bunches of cut herbs averaging fifteen hundred a week and flower and vegetable production smaller and more variable. "We're not holding still, though," Amelia says. "You don't hold still on a farm, but we're holding steady on the things we know from Mom and our own experience how to do."

With the exception of Amelia in her duties as operations director, the same people continue to perform the same tasks as before. Amelia runs the dairy and shares milking and sanitation chores with employees and Ben. She is also chief cheese maker, assisted by an employee. Denny and the employees plan, plant, harvest, and bunch the horticultural crops and handle the composting. And except for short periods when he is in California or on other travels, Denny also makes deliveries each Tuesday and Friday. He maintains Pure Luck's accounts with Whole Foods, Central Markets in Austin, Dallas, and Houston (which supplanted the regular H-E-B stores that previously provided venues for Pure Luck), and several restaurants in Austin and Houston. (Products for Dallas and Houston clients are sent through a parcel service.) Gitana manages the payroll, issues invoices, and receives payments, and Amelia writes checks for farm operation expenses. Twenty-year-old Claire, who lives in Austin and holds a part-time job there, also works part time at the farm, mainly packing cheese. Hope helps with the goats, worming and feeding them.

With so many people working together, and many of them living at the farm, they are all affected when something goes wrong. "We're dependent on each other and on pretty much the same things," Amelia says. "Like the water well. We have one well here, and our houses and the dairy and the crops all depend on it. So when there's a problem, whoever notices calls somebody else. Like I might call Hope or my dairy helper, and say, 'Hey, I have all this shampoo in my hair and I'm all soaped up, and my water just went off! What's going on where you are? Have you heard from anybody else?'"

Sara was always the one to see that most crises like this got resolved, and that longer-term maintenance issues were also addressed. "She was our director of everything," Amelia says. "She kept her eye on our equipment and buildings and would figure out whether we should fix or replace some particular part of something or replace the whole thing, get a whole new piece of equipment. I mean, she would do the reading and the research and make the phone calls and, if need be, go check something out." Now Amelia does these things. "It's part of being el jefe," she says. "I don't get to just put something on a list anymore and leave it at that. It's a big world!"

The world feels bigger than before to Hope, too, but not entirely for the same reasons. She attends school in Dripping Springs, as all her sisters did, and with only two years of high school left she is thinking more and more about college. She is not sure what she wants to

Goats drink whey outside the Pure Luck milking parlor.

study, but she thinks she would like to go out of state, maybe somewhere she has never been before, someplace completely new. Yet the thought of leaving home makes her anxious. She dreads how much she will miss her sisters and the farm. "I love living here," Hope says. "I love this land and the goats and the cheese and going outside to get just about any kind of food for dinner."

Growing up, Hope spent most of her time outdoors. "We went to the creek a lot and swam, and just ran through the fields, and played in the dairy, jumping in the hay," she says. "I think I realized pretty young how good I had it. I think I knew not everybody got to do what we did." The only thing she never had that made her envy other people was a horse. "I was horse crazy," she says, "and so was my best friend, Katie, and we pretended the goats were horses, colts. That was one of our favorite things!"

Amelia points out that, not only has Hope never lived anywhere else, she was born in the room that has always been her bedroom, and just being in it somehow nurtures and steadies her. Sounding like an older sister and a mother both, Amelia then addresses Hope's ambivalence about living elsewhere. "Hope knows this is home and knows she can leave and come back, and all of us and everything on this farm will be here," Amelia says, looking at Hope and no one else. "And Hope knows that her bedroom will always be her bedroom."

Amelia has been making an adult life for herself at Pure Luck for almost ten years, since 1997, when at the age of twenty she eased into Sara's cheese business. Amelia had a job in Austin at the time, doing computer systems work with her father, who owns and operates a business there. She wasn't sure how much she would like cheese making, or the pay cut it entailed, so she tested it, working with Sara a couple of days a week and in Austin three days a week. "We were still small scale with the cheese then, and Mom and I had a really good time," Amelia says. "And after eight months or so, I stopped working in town and went into cheese full time."

She loves making cheese. "It's interesting yet relaxing," she says. "You think and calculate with your eyes."

Though she chose cheese making over computers, Amelia is glad she worked in the computer world for a while. It helped her realize how much she wanted to be outside and made farm work seem like a real choice, not something inevitable or coerced. "Working in Austin, sitting in an air-conditioned office all day, I kept wishing to be outside. By noon, I'd be counting down till I could come home and go outside," she says. "I'd never spent so many hours, so many days at a time cooped up like that. Growing up, we were almost always outside. We played all over the place, and we often ate outside, on picnic tables. And even inside, we felt the outside because we kept the windows open, except just when it was really cold, and we didn't have a TV or air conditioning. We had fans and the breeze."

Another link between the inside of their house and the outside was a small population of spiders. "Mom taught us that spiders do good things," she explains, smiling broadly, "and showed us how to identify and avoid the venomous ones, like scorpions. And we're organic, right? We don't spray chemical poisons on bugs! So we've always had these interesting spiders inside, trapping roaches in their webs, and trapping flies and other insects you really don't want."

Amelia has had her own house since 2001, when she bought a 600-square-foot bungalow in Austin and moved it to Pure Luck. "Spiders have always been welcome in my house, too," she notes. Besides spiders, she and Ben also share the house with Paco, a parrot who was born a couple of months before Hope, and a young mutt named Zoe.

Amelia and Ben met in 1997, when they were in high school, and married in 2006. Ben has a degree in photo communications from St. Edward's University but joined the Pure Luck team as an employee a year or so before he and Amelia married. He mainly helps milk and does other chores with goats. He likes the work, Amelia says, and he also likes the way you can play on a farm. He is

Amelia Sweethardt and Hope Bolton in a field of flowers and herbs at Pure Luck.

building a kind of treehouse. Still in the framing stage, it's actually a one-room house on stilts, nestled among some live oaks and cedars not far from their bungalow. Amelia walks us past it on a tour that takes us from Hope and Denny's house to the goat yard, then past the dairy barn and on to the compost area, then the bungalow, and finally down to the plots of herbs and flowers.

Only a few days after the solstice, summer is doing its work. Zinnias are blooming thigh high, in red, yellow, pink, and purple. Big-faced sunflowers bloom shoulder high, and the newest shoots of long-established rosemary bushes rise over our heads. The marigolds have yet to flower, but their leaves are dense and buds are coming on. Basil is the lowest-growing thing around, yet it too is dense. And in places so are the weeds, Amelia points out. "We don't like weedy patches," she says, "but, you know, weeds are natural. And we're not a farm that's generally open to the public, so it doesn't pay us to spend time maintaining things for the sake of appearances."

Walking between a couple of especially weedy rows, Amelia stops and uncovers melon vines with one of her feet. "In case you didn't notice these," she says, chuckling, "or did notice and wondered just how weedy we let things get." The melons are not a cash crop, she explains, but something an employee has planted just for fun, just for himself and his family and anybody on the farm who wants some to eat. "Whether we're playing or working here," Amelia says, becoming reflective as she guides us along, "it's like I was telling Gitana, and I think she gets it—this farm, with all the people involved in it, is like a river, really. It keeps moving. One person does something, or doesn't do something, picks something up for a while, then drops it, and so on, and the farm keeps moving. It's bigger than all of us and just keeps on going. And the goats! The goats are always making milk, and that in itself is a force."

Amelia feels overwhelmed sometimes, trying to keep up with everyone and all that goes on, but by and large she likes the pace and loves their productivity. "And I don't mean just our economic success," she says, "but the quality of what we produce, and the beauty."

Contact information

Pure Luck Farm and Dairy
101 Twin Oaks Trail
Dripping Springs, TX 78620

Telephone: 512-858-7034
E-mail: pureluck@purelucktexas.com
Web: www.purelucktexas.com

Full Quiver Farm & Dairy

How the Sams Family Saved Its Farm and Sustained a Way of Life

At midday on a Friday, in the middle of a dryer and hotter than normal July, most of the cattle at Full Quiver have sought the deepest shade for grazing. The chickens, a single, small flock of them, forage beneath the roof of their portable wooden shed, and a dozen piglets doze in the leaf-littered dirt of their shaded paddock. Two pet horses, a seventeen-year-old white quarter horse named Molly and a five-year-old pinto mare named Princess, claim their own strips of shade, venturing into full sun only briefly. Just as sensibly, Mike and Debbie Sams, together with sons Joshua, eighteen, and Levi, sixteen, are working in the air-conditioned comfort of their cheese plant, a small room brightened by white walls, fluorescent ceiling lights, and five windows.

Standing at stainless steel tables over large stainless steel bowls, the four Samses are plying batches of mozzarella with gloved hands, shaping it into balls and braids. Mike, big, dark, tall, and talkative, wears a baseball hat and a blue plastic apron over a blue shirt and jeans. Debbie, who is medium-size and has fair skin and dark hair pulled into a bun, wears the same type of apron over a floral print pinafore and blue dress. A bonnet of pale pink gauze covers the back of her head. Like Mike, Debbie talks readily and without faltering either in conversation or her work. Joshua and Levi, dark like their father but not as tall or big, are much more reserved than their parents. Instead of aprons, they are wearing white lab coats, a garb that, even though they are also wearing baseball hats, somehow highlights their reserve and the single-minded concentration they seem to give to their task.

Ordinarily, the two youngest children, Hannah, fourteen, and Seth, eleven, also participate in cheese making, but today they are visiting their grandmother, Debbie's mother, who lives in the area. Mike and Debbie, both in their early fifties, are the parents of nine children, ranging in age from eleven to thirty-six. The five oldest have married and left the farm, and the four youngest remain.

Since 2002, the Sams family's main commercial product has been cheese. On the pastures of their sixty-three-acre farm, located near Kemp, about fifty miles southeast of Dallas, they maintain a herd of thirty Holsteins and Holstein-Jersey crosses. Milking from twenty to twenty-five during any given period, the family makes mozzarella, cream cheese spreads, and several aged hard cheeses. They sell their cheese wholesale to the Wheatsville Co-op in Austin and to Central Market and Whole Foods in both Austin and Dallas, and they retail at the Sunset Valley Farmers' Market in Austin. On a lesser but still economically significant scale, they also raise chickens, hogs, and beef cattle on pasture and sell the meat at Sunset Valley.

The Full Quiver cheese plant.

The terrain of the farm is mostly flat and open and lies within the post oak belt of east-central Texas. Mike describes the soil as "post oak sand," or sandy loam. "Good soil," he says. Native prairie grasses spring up in the pastures now and then, but cultivated nonnative species predominate. Bermuda grass is the main warm-season perennial grass, into which Mike sows annual Sudan grass during spring and annual rye grass in the winter, trying to ensure continual forage. They fertilize the pastures with manure from the milking barn, adding to that which the animals naturally drop. During especially dry periods, if their pastures do not produce enough grass, they buy alfalfa hay to feed the cows and beeves. "We're a grass-based dairy and grass-based meat farm. That's what we strive for," Mike explains. "But when there's no rain and hardly any grass, you've got to feed them something!"

Pecan trees and oaks—mainly post oak, bur, and red—grow along interior cross-fences and perimeter fences. And in one pasture, all by itself stands a single mesquite. The Samses are from West Texas and love mesquite, though they don't permit any more to take root, knowing how easily mesquite can spread and ruin good grassland. "But they're so beautiful in the spring, with all those little gold blossoms," Mike says, "we have to have just one."

Plenty of pasture and relatively few trees are the features the Samses had in mind when they found the place, in 1993. "We wanted some trees for shade but preferably out of the way so we could grow grass," Mike says. By then, they had been small-scale, wholesale dairy farmers for eleven years, eight of those years near Abilene and three in Dublin, Georgia, where they had moved after becoming Mennonites, to be part of a Mennonite community made up largely of dairy farmers. But the pull of family and their abiding sense of Texas as home soon drew them back, and they chose the Kemp area instead of West Texas because of a Mennonite community at Scurry. Though the Scurry group is mainly made up of woodworkers, not farmers, it welcomed the Samses and the wholesale dairying the family envisioned as their future.

"When we came here, the only farming we knew was growing grass and pasturing cows and shipping milk commercially," Debbie says. "We sold to Dairy Farmers of America, and a big truck would come every few days and pick up the milk and pull out with it. Some of it went to ice cream makers, some to bottlers and distributors." They were generally milking at least 150 cows then, out of a herd that sometimes numbered up to 250, so they had to be able to grow a lot of grass.

And grass, in fact, was all the place had to offer. There was no house, no barn, no building of any kind. But because Mike is a skilled carpenter, welder, plumber, and electrician and has taught many of these skills to their children, the lack did not deter the family. They bought a four-room frame house in Athens, Texas, moved it to the farm, and added two rooms onto the back. Then they built a milking facility, a tan-colored metal structure that at 24 by 200 feet, including a 24- by 50-foot milking parlor, is at most a fourth the size of some of the milking facilities at feedlot dairies. They also built a metal tractor-and-implement barn about the same total size as the milking facility. Nine years later, in 2002, they constructed the small cheese room—at 18 by 22 feet, it is smaller than most single-car garages—and an adjoining cold-storage room which, at 12 by 20 feet, is even smaller. All of these buildings are clustered near the farm's center, on either side of a gravel drive that leads from the county road into the farm.

Tomorrow is Saturday, a long day of marketing in Austin. The mozzarella they are making will be only about sixteen hours old at nine A.M., when the first customers at the Sunset Valley Farmers' Market get their hands on it. Customers will also find a variety of cream cheese spreads, made only seventy-two hours earlier, on Wednesday, the Samses' other main cheese-making day. Aged, hard cheeses—colby, cheddar, and jack—and meats will also be offered. At no other venue is the whole array of Full Quiver products available, and though profitable for the farm the market is also costly in terms of family labor and time. Joshua and Levi take turns accompanying Mike to the market, alternating Saturdays, and sometimes Seth also joins the crew, pitching in at both booths. Debbie and Hannah occasionally go also, but not often.

"We've got two booths, one for cheese and one for meat," Mike explains. "We're side by side. Josh or Levi, whichever one's there, runs the meat booth, and I run the cheese booth. We pull out of here by four A.M. every Saturday and get into Austin by seven. Before the market opens, we go to two Whole Foods stores, two Central Markets, and the Wheatsville Co-op. By eight fifteen or eight thirty, we're setting up at the market and selling from nine till two. And then we break our canopies and tables down and load back up. If traffic's not too bad, we're home by six."

The trip to Austin is two hundred miles each way. To simplify Saturday's workload and save time, Mike could skip going to individual Whole Foods and Central Market stores and instead rely on each company's distribution system to get their cheese on the shelves. But he chooses to go the extra miles because he wants to ensure that their products are presented well and in a timely way. "We want our cheeses to be at their best when people buy them so they'll know just how good they are and keep buying them," he says. "If things are held too long before getting out on the shelves and aren't fresh, or if they're put where people won't much notice them, it'll hurt our business."

For the same reasons, Mike drives to Dallas every Thursday, about a hundred-mile round trip, and delivers cheese to individual Whole Foods and Central Market stores there. Although the better part of their income is from their wholesale accounts with these stores and the Samses like their relationships with store associates, they enjoy the farmers' market much more. "We know we're giving our customers products that are in peak condition, and we can talk with them and tell them just what all goes into what they're getting. And then, too, we hear and learn from them what they like, and that helps us stay on our toes," Mike says. "Plus we get more for our product."

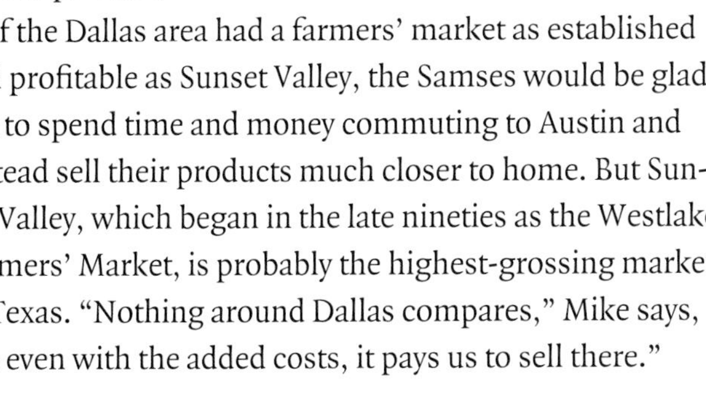

If the Dallas area had a farmers' market as established and profitable as Sunset Valley, the Samses would be glad not to spend time and money commuting to Austin and instead sell their products much closer to home. But Sunset Valley, which began in the late nineties as the Westlake Farmers' Market, is probably the highest-grossing market in Texas. "Nothing around Dallas compares," Mike says, "so even with the added costs, it pays us to sell there."

Mike Sams extracting mozzarella curds from whey.

Switching from wholesale milk production to cheese making in 2002 resulted from economic stress the

Sams experienced throughout the nineties as the dairy industry, like other livestock industries, consolidated into corporate-owned, confined-animal feedlot operations. Rather than grazing on grass, cows in feedlot dairies are densely confined in huge barns with concrete or dirt floors. There they are fed mostly grain and given synthetic hormones, to increase the volume of milk and ensure a more consistent, predictable supply than grass-fed cows produce. This process of consolidating many small grass-based dairies into relatively fewer large feedlots drove wholesale prices for a gallon of milk almost relentlessly downward and forced thousands of small dairy farmers out of business.

Hoping prices might somehow rise, the Samses held on by starting a side business to supplement their dwindling dairy income. Working with their oldest son, Josiah, who still lived at home then, Mike began building and selling storage sheds. This helped but not enough. As the decade ended and milk prices fell even further, they were struggling more than ever to pay their mortgage and other expenses. One of the few things they did not have to worry about, they note, was costly medical insurance. As Mennonites, they participate in a "brotherhood assistance" program. "We all help cover each other's costs as the needs arise," Debbie explains. "And this saves us and everybody a lot of money. But even so, over a period of about twelve years we kept getting in a deeper and deeper hole, and when you can no longer see the top of the hole from the bottom, you think, 'Well, we've got to do something different or we'll have to sell the farm.'"

During the course of searching for a solution, they read several books by Joel Salatin, the noted Virginia farmer who with his family raises a variety of meat animals and laying hens on pasture. Debbie loves chickens, and Salatin's *Pastured Poultry Profits* particularly excited her. "I said to Mike, 'Oh, we could do that! We could do pastured poultry!' And Mike said, 'Well, you know, we have all these dairy cows on hand. Wouldn't it make more sense to do something with milk? Which we already have?'"

Remembering that conversation, both Mike and Debbie laugh. "Well," Mike says, "we were fixing to make a last-ditch effort to save the farm. What else could I say? There wasn't any time or money to risk on something new and big."

They set their sights on cheese. Debbie was used to making it for the family, and they felt that, without too much difficulty, they could build on her knowledge and skill, but they knew nothing about possible markets. They sounded out their friends and acquaintances for tips, and someone suggested Whole Foods, which they had heard of but never been to. Figuring it was worth a look, they went to a Dallas store and liked what they saw. They talked to someone in the cheese department about their products, and she referred them to one of Whole Foods' regional buyers in Austin. Debbie called the buyer, and the buyer told her Whole Foods was opening a new Dallas store. She asked Debbie to meet her there on a certain date and bring samples. "So we took her some mozzarella and a few cream cheese spreads, and she said, 'Whenever you're ready, let's go!'" Debbie recounts. "Which was great, of course. But we hadn't built our facility yet. We were waiting to see if we had a way to sell."

Mike and Debbie got busy working with the state health department then, to learn what facilities and procedures were required for commercial cheese production. "We asked them what were the bare bones for doing what we wanted to do," Debbie says, "and they were helpful, and once we understood the bare bones, we borrowed some money and started building." Initially, the Samses could afford to build only the cheese room and not also a cold-storage room. This meant

Joshua and Levi Sams making mozzarella rounds.

using several big, old refrigerators they already had until their cheese production and sales increased enough to allow for adding on the cold-storage area.

Though small, the cheese room is ample for worktables, sinks, supply cabinets, and a pasteurizer, the largest piece of equipment in the room, and the only machine. Like the goat milk cheese of Pure Luck (profiled in the previous chapter), Full Quiver's cheeses are handmade exclusively from milk produced at the farm. In the professional cheese world, this means they can be designated as both artisanal and farmstead products. Though the Samses do not enter formal competitions, Debbie expresses a quiet pride in these distinctions. "We do everything by hand with just our milk, in just this little room," she says, "just our family, the six of us here, and it works fine."

Since 2002, they have given the shed-building business entirely to Josiah, who lives nearby with his wife and two children, and increased their weekly sales of cheese from 300 to about 1,200 pounds. One market venue led to another. Their success at Whole Foods made the decision to add Central Market easy, Debbie explains, though adding Sunset Valley Farmers' Market was something they resisted for almost a year. They didn't know there was any such thing as professionally managed farmers' markets or have any idea of their increasing popularity and profitability. "We thought of farmers' markets as flea markets," Debbie says, "and imagined standing all day in the sun for maybe fifty dollars. So we balked." But the market managers persisted, telling the Samses they would come home with good money. In August of 2005, they decided to give it a try and have been going ever since.

Not only has Sunset Valley proved a good outlet for their cheese, but it serves as the sole venue for most of the meat products they began developing soon after joining the market. The Samses had long known about the nutritional benefits of eating dairy products, eggs, and meat from pastured animals rather than confined: less overall fat and less saturated fat; more beta-carotene, folic acid, and vitamin E; and a better proportion of omega-3 fatty acids, which aid in preventing heart disease and strengthen the immune system. Reading about nutrition had taught them that. And they knew how good grass-based animal products taste from years of producing them just for their family. But they did not know that so many urban people in their part of Texas were also savvy about these things and would stand in line at a market

Joshua and Levi Sams in the Full Quiver milking barn.

to make purchases, or even just to put their names on a waiting list. "It took customers and meat farmers we met at Sunset Valley to open our eyes," Debbie says. "And so my pastured poultry idea kind of worked out, after all, as a good little sideline for us!"

The Samses have not sought organic certification, though they meet or exceed the standards. They give no hormones to any of their animals, nor, except in cases of illness, do they administer antibiotics, and they do not apply pesticides or herbicides to their pastures. One indicator of the soundness of their practices, Mike points out, is the health of their animals. Some of their oldest dairy cows are thirteen to fifteen years old and still calving and producing milk, compared with an average of about eleven productive years for dairy cows in general. Mike describes their grazing methods with all their animals as rotational but not intensively so. This contrasts with Salatin and with emulators like the families of Windy Meadows Farm and Rehoboth Ranch (profiled earlier), who concentrate a larger number of animals in smaller paddocks than the Samses do and move the animals more frequently. However, since the national organic rules for livestock allow for grain-based diets and require only "access to pasture" and not that the animals actually gain it, the Samses can claim to be going "beyond organic," just as Salatin and those emulating him more closely do.

Having customers they can talk with about their methods and the issues involved, the Samses find formal certification irrelevant. "There's a lot of paperwork and expense involved to get certified, and things are working great the way they are. We're selling all the cheese we can make and all our meat," Mike explains. "If certification would give us more market or more of a premium and we needed those things, then it would be a marketing tool we'd use. But so far it's not, so why do it?"

The Samses raise about a thousand meat chickens a year, and eleven-year-old Seth takes care of them. For processing, the family takes the broilers to Windy Meadows Farm, where the Hale family has a state-licensed facility. Eighteen-year-old Joshua handles the beef steers and pigs, finishing about fifty beeves a year and about sixty hogs. These animals are processed at Burgundy Pasture Beef, in Grandview, a family-owned and operated ranch, licensed processing facility, and store, which also buys some of Joshua's meats and sells it under their label. An Angus-Holstein cross is the breed they prefer for beef, because it produces tasty cuts and they wanted to use what they already had, and for pork they like Chestershire-Hampshire pigs. Joshua supplements the pigs' forage with whey left over from cheese making, which results in mild-tasting pork that a lot of people prefer, Mike notes.

"Joshua's getting set up to make a living in ranching," Mike says. "He's getting good retail prices for his cuts, plus we pay him and all of our children for the work they do. We want them to make enough money that they're satisfied to stay here on the farm and learn things that will serve them now and later on, if that's what they want."

Debbie keeps a vegetable garden, which provides the family with most of its vegetables year round, either fresh or preserved. This July day, the garden, a large plot near the house, lies fallow, resting after a bounteous spring and early summer. Debbie has cured all the onions they didn't already eat—a bag is hanging from a tree—and put up all the green beans, peas, squash, and tomatoes they will need until next spring, when the garden is bearing again. Fruit, plus any vegetables she doesn't grow that the family might want, they buy or barter from neighbors and farmers at Sunset Valley.

Debbie used to can everything she put up, but she has come to prefer freezing because she thinks it generally preserves more nutrients. Some vegetables, however, especially beets, cucumbers, and cabbage, she preserves through a process called lacto-fermentation, which involves whey. A farmers' market customer introduced her to this method, and to the work of Sally Fallon, perhaps the best-known proponent of such traditional techniques and head of the Weston A. Price Foundation, an alternative nutrition organization. Debbie bought Fallon's cookbook, *Nourishing Traditions,* after listening to a tape a customer gave them of Fallon lecturing. Recipes for all her lacto-fermented vegetables come from it. "We love how our customers introduce us to new things!" she says. "We didn't know what we were missing, but now we're hooked on lacto-fermented veggies."

Debbie also keeps their family supplied with yogurt, kefir, ice cream, and butter, and she bakes all their breads with wheat she grinds herself. The wheat comes from Montana, but she buys it from a bakery in Kemp operated by a family in their church. "They sell certain items in bulk, and besides wheat I buy other grains and raw sugar and good salt and oatmeal." Debbie says. "But there's not much we need to buy from them or anyone. And we like that. We really enjoy being as self-sufficient as we can."

Though Debbie enjoys her roles in the family's commercial food production—helping make cheese, helping tend the animals—she most enjoys preparing food for their family and keeping house. Yet everything she does is of a piece, and it all gives her pleasure. "We live the way I've always wanted to live," she says. "Since before I can remember, this is the life I wanted."

Two farms in her earliest childhood nurtured this desire. One belonged to her grandparents, the other to her aunt. The farms were just down the road from each other in Farmersville, not far from Garland, the Dallas suburb where Debbie, born in 1953, lived with her parents and younger sister the first eight years of her life. Debbie's grandparents and aunt also lived in Garland, but they worked at their farms on weekends and other times, especially in the summer. Debbie's father worked in a bread factory and her mother was a homemaker, and neither was particularly interested in the farms, nor was Debbie's sister. But Debbie loved being at the farms, and every Friday afternoon her grandparents or her aunt, together with her son, Debbie's cousin, picked her up on their way out.

"I couldn't imagine anything greater," Debbie says. "I would go out into the cotton fields and try to help chop weeds even though I was just little. And I'd get hot and sweaty and dirty, and I thought that was the grandest thing! And they had cows, just a few, and milked them by hand, and I would sit in the corner of the stall while one of the grown-ups milked, and then I'd sit by the big milk can and hold it steady and keep the flies off when they poured buckets of milk into it. And by dark, I'd be just filthy dirty, but there was no running water in the house—we drew water from a well and carried it to the house—and we took baths out on the porch in a big metal washtub. And we slept with all the windows open. There was no air conditioning. The house was always open during the warm months, and I just loved listening to the crickets and frogs at night. And I thought all that was living!"

Not that she didn't love daily life with her parents and sister, Debbie notes. "We bought a brand new house with my daddy's veteran loan, just a tiny little thing but brand new and comfortable, and my mother kept it neat and clean," Debbie says, "and she dressed me in frilly little dresses and put me in the bathtub and scrubbed me every day and shampooed my hair, and I loved that too."

Debbie Sams in the Full Quiver farmhouse.

Even so, life in Garland paled beside life in Farmersville. "Everything about being there was wonderful to me," she says. "I mean, I can still see the golden sun streaming through the sides of the barn onto the floor, and that just thrills me to death and I don't know why!"

Mike had no such connection to farming while he was growing up. "I'm a city boy!" he jokes. "Used to be, anyway!" The fifth of six children, he was born in Amarillo the same year Debbie was born, in 1953. His mother was a homemaker and his father worked as a warehouseman for Furr's, the cafeteria and grocery chain. When Mike was eight his family moved to Lubbock, and so did Debbie's, because both their fathers received transfers. They met in junior high, at the age of twelve, and married in 1969, when they were sixteen, with a couple of years in high school left to go. By the time they graduated, their first child, a daughter, was a year old, and Mike went to work full time for Furr's. Debbie took care of their growing family—four daughters by the end of the seventies—and though they lived in various towns in the Panhandle and for a while in Denver, wherever Mike was transferred, she kept a vegetable garden and, when possible, also chickens for eggs and meat.

Mike had grown up in the Church of Christ, Debbie in the Southern Baptist, but early in their marriage they were drawn to the Assembly of God and joined a charismatic congregation. They found not just the beliefs congenial but even more the lifestyles of most of the people themselves. "This was the seventies, the hippie days, when so many people, not just hippies, were coming back to the earth, back to the land," Debbie says. "And to me that was the charismatic church—mothers staying home with their children and having gardens."

As the seventies ended, however, Debbie and Mike felt a gap growing between their own increasingly traditional lifestyle and that of their fellow charismatics. "Most of the ladies were getting jobs and going right back to the modern ways," Debbie says, "And Mike and I started thinking we didn't want our girls to grow up and marry boys brought up that way, and we decided to make some changes."

Mike stopped working for Furr's, and they bought an old two-story store in the rural community of View, outside of Abilene. The store included a post office and occupied the lower floor, and the family lived upstairs. Running the store and post office, the Samses soon began receiving offers to do odd jobs at local dairies. "There were quite a few widows who didn't have children around to help them with their cows," Mike says, "and it didn't take both of us to run the store and post office the whole time, so I helped them out and loved it. And our four daughters went with me quite a bit, and they loved it, too."

Before long, one of the dairy owners offered Mike a full-time job. It seemed the perfect opportunity except that, if they kept the store, the commuting distance would take Mike away from home much more than they wanted. So they sold the store and bought twenty acres with a house near the dairy. Feeling blessed to be living with their four children on their own farm, and hoping for more children, they named the farm Full Quiver and carried the name with them when they moved to Georgia and then to Kemp. It comes from Psalm 127, which celebrates children as a heritage of God and a source of strength and joy: "As arrows are in the hand of a mighty man, so are children of thy youth. Happy is the man that hath his quiver full of them."

About this time, they also decided to school their children at home. Three of their four daughters were in school then, the oldest in eighth grade, and Mike and

Debbie took all of them out. "We felt it would make our family even closer," Debbie explains. "The children didn't have to spend time riding the bus anymore. They could get their schoolwork done early enough in the day to be outside a lot. They could help with the chickens and the garden. But mainly they played. They rode their bicycles for hours at a time. We knew everybody around, and they could stop at any house and visit and do things with our neighbors."

In searching for a course of study, Mike and Debbie found that the Mennonite curriculum seemed both academically sound and in harmony with their values and way of life, and so they chose it, a choice that would gradually lead them into the Mennonite Church. "We tried another curriculum at first, but the examples in the lessons just didn't apply to the things we do. I remember some lesson in science about conserving energy by opening the dishwasher and letting air dry the dishes instead of using the dry cycle! Well, we've never had a dishwasher and don't plan on getting one!" Debbie says, laughing. "Examples like that make children want things we think aren't worth wanting. In the Mennonite curriculum, even the story problems in math often have to do with farm animals and farming activities."

Debbie says their children have always enjoyed reading and studying. "And we read together out loud as a family in the evenings, especially in winter, when the days are shorter, and on Sundays. We've always got a book going. We take turns picking what it is."

"And we've never had a TV or a computer," Mike notes.

"And when kids don't have a TV or a computer, they'll read anything," Debbie adds, "and love it so much they won't want to give it up. We've always just had to pry our children away from their books."

They recently discovered the nineteenth-century British novelist G. A. Henty, a prolific author who wrote mostly historical fiction for children, and are excited about working their way through his long list. "The first couple of books we got by him, one on the California gold rush, and one on the Civil War," Debbie says, "our children read so fast on their own that we didn't have a chance to read them as a family!"

They schooled their children at home until 1990, when they joined the Mennonite community in Dublin, Georgia. Since then, both in Georgia and at Kemp, they have sent their children to the local Mennonite school. All of their children remain in the Mennonite fold. The two oldest daughters married Mennonites from Pennsylvania, one a farmer, one an accountant, and live there. The two next oldest daughters, like their oldest son, married Mennonites in the Kemp and Scurry area and live nearby. One of these sons-in-law works in construction, and the other teaches school. The married children have so far given Mike and Debbie thirteen grandchildren, with two more expected. And even though only one of the five couples farms for a living, all live rural lives.

Late in the afternoon, though the heat of the day persists, our conversation comes to a cool, sweet end: Debbie serves ice cream she made earlier in the day. Leading us inside the house, she advises us that they have air conditioning but don't like living in it and hardly ever use it. "It's not on today," she says. "I hope you can be comfortable." And we are. The house has plenty of windows, some of them double. They're all raised, and ceiling fans are turning. The cross ventilation in the main room—the kitchen on one end, the living room on the other—couldn't be better. Now and then, the breeze picks up enough to stir the curtains. And it also brings us the sound of a braying donkey. Everyone laughs.

Cows rest in the shade at Full Quiver.

"He belongs to one of my grandsons," Debbie says. "We never had one before. I didn't know donkeys made all that racket!"

Debbie and Mike wish that all their children and grandchildren lived nearby, but with everybody well and happy, they take heart in that. Occasionally Debbie manages to fly to Pennsylvania and visit for a few days, but more often the Pennsylvania children and grandchildren come to Kemp. Just last week, they came. "The oldest grandchild is eleven, the same age as our youngest, Seth," Debbie explains, "so we had all these little children playing here, plus our fourteen-year-old Hannah and of course Joshua and Levi joining in, and everybody had the best time." Joshua and Levi got some old sheets and resurrected a tepee they used to play in, and Hannah made little headbands with chicken feathers sticking up, and Seth made bows and arrows out of PVC pipe and bailing twine, and everybody played Indians. "They even hooked a sled-like contraption onto our horse Molly," Debbie says, "just like our grown-up children used to do, and took turns scooting around the pastures like an Indian on a sled!"

Debbie and Mike savor these times, but perhaps no more than they do most days of their lives. "We get to make a living doing what we love. Every day we live what we love, and we feel so blessed and grateful, because you've got to have some kind of business to survive in this world. You have to make money some way," Debbie reflects. "But even if farming weren't our business, even if Mike had to put on a suit and drive to work in Dallas every day, I'd still be milking a cow and making butter and cheese and digging in my garden and drying our clothes on the line. And we'd still be raising laying hens and meat animals, and all of us would still be working together and playing together and still be living what we love."

Contact information

Michael and Debbie Sams
Full Quiver Farm
6238 FM 3396
Kemp, TX 75143

Telephone: 903-498-3884

Index